AF876706

PURCHASED FROM
MULTNOMAH COUNTY LIBRARY
TITLE WAVE BOOKSTORE

QUEST FOR A NORTHERN AIR ROUTE

QUEST FOR A NORTHERN AIR ROUTE

by Alexander Forbes

HARVARD UNIVERSITY PRESS · CAMBRIDGE · 1953

LIBRARY ASSOCIATION OF PORTLAND, ORE.

SECOND PRINTING

COPYRIGHT, 1953, BY THE PRESIDENT AND FELLOWS

OF HARVARD COLLEGE

LIBRARY OF CONGRESS CATALOG CARD NUMBER 52-12262

PRINTED IN THE UNITED STATES OF AMERICA

To the memory of

CHARLES JOSEPH HUBBARD

a loyal friend who opened a new era in forecasting weather for the safety of airborne commerce over the northern ocean and gave his life in the far North, greatly serving his country and all mankind

MAR 23 1954

FOREWORD

It would be difficult to find another man so versatile as the author of this book. Alexander Forbes, internationally known as physiologist, is also a familiar figure on the ski slopes of New England, on the skating rinks, and in sailboats anywhere you can name between Long Island Sound and Baffin Land. Also, one may add, in the Baltic and the Aegean, where I made a memorable cruise with him in 1934. From seagoing Forbeses he inherited a share in Naushon, that fabulous island off the New England coast, and a deep love of seafaring. As a practical navigator with sextant or dead-reckoning, few professionals are his equal. He flies a plane and sails his own ketch, at the age of seventy. His hardihood and indifference to creature comforts are the constant astonishment of his friends. Two years ago he sailed a small open lifeboat from Boston to Mount Desert just for the fun of it; the young lieutenant commander who shared the same tarpaulin and the exiguous meals, admitted that it was pretty rugged but that Dr. Forbes's company made the cruise a wonderful experience.

During World War I, as lieutenant (jg) in the Naval Reserve, Dr. Forbes was put to work installing and testing radio direction-finders in destroyers. His inventiveness was again placed at the disposal of his country in World War II. On one of his yachting cruises, in the Blue-nose schooner Ramah, *he had organized a survey of a hitherto inaccurately charted section of the Labrador coast. That qualified him for the adventure into the Arctic that is here related. There was no compulsion, at the age of sixty, to return to active duty in the Navy, but that is what he did; his mission to select airfields in the Canadian North was performed with gusto, as the reader of this book will presently ascertain.*

November 1952 SAMUEL ELIOT MORISON

PREFACE

Far to the north of our country lies an archipelago of gigantic islands comprising an area nearly half as large as the United States, separated from continental America by wide bays and narrow straits, bounded on the east by Baffin Bay and the adjacent shores of Greenland and on the north by the Arctic Ocean. A century ago the famous Franklin expedition was lost in the heart of that vast archipelago, and in subsequent decades many explorers went in search of survivors or vestiges that would tell the story of their fate. Consequently the archipelago is remarkably well mapped for so large and inaccessible a region, but still its intricate coastlines are so extensive and so remote that their geography is but crudely delineated, while the interior is largely unknown. This great land of islands might have been of vital importance in World War II; it may yet be a key to the prevention of World War III.

Whether as a locus of defensive outposts or as a meeting ground for friendly coöperation in polar research with our European neighbors, the lands that bound the North Polar Sea are of major concern to coming generations.

Many persons engaged in the war effort visited this northern area between 1940 and 1945 on missions pertaining to air transport and weather forecasting. My part in these missions, though small, was unique in the successive contacts with various phases of the problem, and the experiences I enjoyed seemed to me worth recording.

Grateful acknowledgments are due to John T. Rowland, Samuel E. Morison, Bliss Perry, William S. Carlson, and E. C. Sweeny, all of whom read the manuscript and offered constructive and helpful criticism. Especial thanks are due to Barbara Burleigh Howe, whose skillful editing put a spark of life into the narrative and transformed a dull recital of details into something acceptable to the publisher. I wish also to express my thanks to R. L. Williams for his cartographic skill in preparing the illustrative maps, to the U. S. Navy and the U. S. Army for furnishing copies of their photographs and permitting their publication, and to the *Harvard Alumni Bulletin* for permission to reprint material published therein.

November 1952 ALEXANDER FORBES

CONTENTS

-1941-

-1942-

-1943-

ILLUSTRATIONS

MAPS AND DIAGRAMS

PART ONE

- 1941 -

1 Northwest River and Goose Bay

Long ago, in talking of the sea, Sir Wilfred Grenfell, the Labrador Doctor, said to me, "If you are fond of cruising, why don't you come to Labrador and map one of the uncharted fiords there?"*

If, when Grenfell had made his suggestion, I could have looked into the future and seen what it was leading to, the vision would have amazed me. The daydream would have led over dizzy heights and through many a deep morass. The lure of the North and the exploration of new techniques for charting by aerial photography led me into a far greater digression from my regular work than I would have dared to undertake knowingly. But then, as if by special planning of Providence, all of this digression suddenly became germane to the issue of World War II.

In the spring of 1941 I was on active duty in the Naval Reserve as Lieutenant Commander in the Medical Corps, attached to the Squantum Naval Air Station while working up the results of some physiological research at Pensacola. In May, a letter from my friend Major Harold B. Willis of the U. S. Army Air Corps, requesting information about Labrador, brought me to Washington armed with photographs taken on our survey ten years earlier. In Washington I learned that a chain of air fields was to be established on the land masses of the far north. These were to

* The survey that Grenfell inspired was made in 1931. The story has been told in the American Geographical Society's publication *Northernmost Labrador Mapped from the Air*. The resulting map, made by O. M. Miller, shows in accurate detail virtually all of Labrador north of Nachvak Fiord; published in 1938, it became the basis of revised Government charts in England, Canada, and the United States.

be placed so close together that fighter planes could reach England under their own power. Within less than a year a few fighter planes in the Royal Air Force had barely saved Britain with a terribly small margin. And now that force was sadly depleted. More were needed as soon as possible. Just a few planes flown quickly to the war zone might turn the scales and save the Allied cause. Long-range bombers were being flown in a single hop from the great airport at Gander Lake, Newfoundland, to Britain. But the fighter planes, unable to span the ocean in this way, were being shipped on freighters and many were sent to the bottom of the sea by German submarines. The losses were appalling and a remedy had to be found. No wonder General Arnold and General Spaatz viewed any way of speeding the flow of planes to Britain as of prime importance. President Roosevelt himself was deeply concerned with the development of the project.

On May 27 Major Willis escorted me to a meeting in the office of General Spaatz. The question of what areas, if any, in northern Labrador would be suitable for an airdrome was discussed. Selected photographs taken on our survey of 1931 were examined intently by the Air Corps officers. Flat areas at the heads of the fiords and farther inland, where we had found the weather much less foggy than along the outer coastline, were of special interest. Reports from other sources revealed that better flying weather was to be expected still farther north in the latitude of Baffin Island. Thus both climate and shorter hops favored a route through the far north. Before leaving Washington I learned that I was to participate in a reconnaissance, soon to be launched, to look for suitable airport sites in Labrador.

One month later, on June 26, I found myself aboard an Army bomber taking off from the Boston airport. As Cape Ann disappeared astern, I marveled at the events which had in so short a time catapulted me from physiological research in Pensacola onto this northbound airship winging its way to Gander, Newfoundland. There I was to meet Captain Elliott Roosevelt, the man in

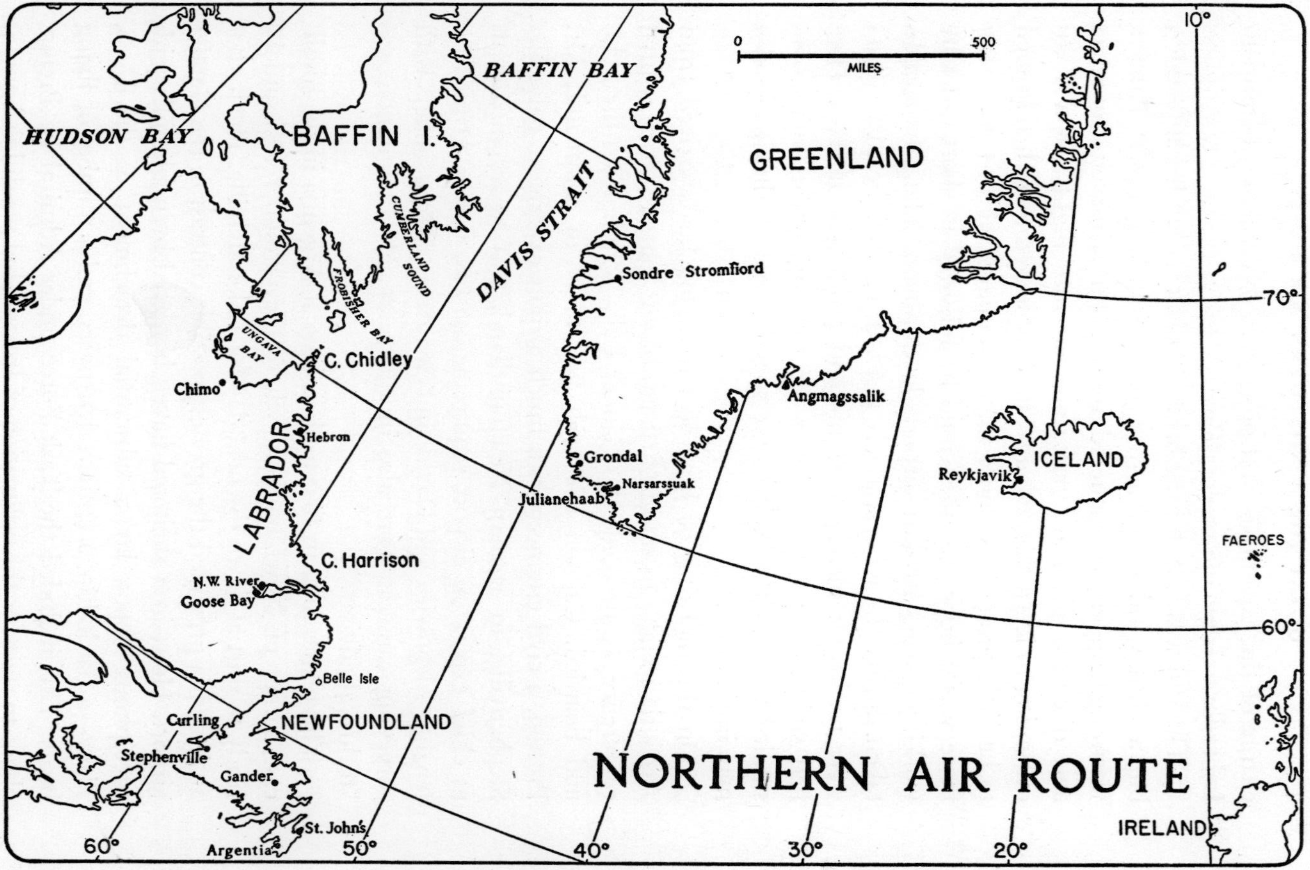

NORTHERN AIR ROUTE
0
500
MILES
HUDSON BAY
BAFFIN I.
BAFFIN BAY
GREENLAND
CUMBERLAND SOUND
FROBISHER BAY
DAVIS STRAIT
Sondre Stromfiord
UNGAVA BAY
C. Chidley
Chimo
Angmagssalik
Hebron
LABRADOR
Grondal
Narsarssuak
Julianehaab
ICELAND
Reykjavik
C. Harrison
N.W. River
Goose Bay
FAEROES
Belle Isle
Curling
NEWFOUNDLAND
Stephenville
Gander
St. John's
Argentia
IRELAND
60°
50°
40°
30°
20°
10°
70°
60°

charge of the expedition, and proceed by Navy patrol bomber to Labrador to begin our explorations.

The sun was still shining brightly when we landed at Gander, but a chill wind swept across the surrounding forest. After a warming supper we gathered in Captain Roosevelt's quarters and discussed plans for our reconnaissance flights. It was encouraging to learn that, whereas in Washington there had seemed to be little prospect of more than reconnaissance this year, now there was a hope that this could be followed in short order by working parties to start airport development. The photographs taken during the comparatively peaceful days of the early 1930's were no longer merely of scenic and geographic interest. They had suddenly become weapons for fighting a war, and as we studied them in the light of the present urgency they took on a new importance.

On the morning of June 30 two flying boats arrived from Argentia to take us on the first lap of our journey. After a final conference we proceeded to Gander Lake, a mile from the field, and boarded. Captain Roosevelt was to be in command; Mr. Preston, a civil engineer with much airport experience gained at Stephenville, in western Newfoundland, was to accompany us as chief expert on airport requirements; two young Army lieutenants, Zienowicz and David, and two Army photographers with surveying cameras were also in the party. My designation was "technical adviser."

The planes took us first to their base at Argentia in the southeast corner of Newfoundland where they would refuel and be ready to start at dawn the next day for Lake Melville. The harbor at Argentia presented the greatest possible contrast to the Gander airport. Here, in a sheltered haven bordered by noble hills with precipitous slopes, lay a substantial detachment of the U. S. Navy—a battleship, a cruiser, destroyers, and a number of flying boats. This included the Task Force under Admiral Le Breton, about to start for the occupation of Iceland, but we didn't know it until our return a few weeks later. We were taken aboard the *Albemarle,* a fine new seaplane tender, where her Commanding Officer and Captain Mullenix welcomed us in the cabin and discussed our project with charts spread out on the table.

Next morning we were roused at 3:50, saw the first glow of dawn over the Newfoundland hills, and were soon aboard the flying boats and headed for the Strait of Belle Isle, separating Newfoundland and Labrador. The sea north of Belle Isle presented a strange picture. A vast ice pack, borne southward by the Labrador current against the north shore of the island, extended seaward for many miles. These ice packs, formed anew each winter, vary greatly in extent and duration from year to year and are wholly different from the tall icebergs that always move southward in a steady procession from the glaciers of Greenland.

Soon we were over Cartwright, 100 miles up the Labrador coast, and then over the mouth of the North River, 50 miles further up the coast, where my photographs of 1935 had revealed an area of level ground. Our one objective was to find a flat place big enough for 5,000-foot runways. We circled over this pebbly estuary and examined it for a few minutes before swinging back on our course to Lake Melville, 30 or 40 miles inland. There several areas were studied and photographed before the planes were set down on the water and anchored off the tiny settlement of Northwest River situated at the west end of the lake. Old friends from the Hudson's Bay Post and the Grenfell Mission located there came out in motorboats and brought us ashore in good time for lunch at the Grenfell Hospital.

It seemed odd, after the secrecy that had surrounded our project in Washington, to find that here it was a matter of common knowledge among the trappers and traders. The Canadians had already been studying the region with airfield development in mind, and the flying boat that brought in their surveyors was already at anchor on Little Lake, close by the settlement.

We were scheduled to start on our quest for an airport site

at 5:30 the next morning, and a motorboat had been engaged for the purpose. Our objective this day was Epinette Point on Lake Melville, 12 miles east of Northwest River, where a large area of swampy but level ground, as seen from the air, appeared to offer possibilities of development. The little cabin motorboat of the Hudson's Bay Company took us across the 10 miles of open water exposed to the full force of the northeaster; she pitched and rolled wildly, but Captain Roosevelt amazed us with his ability to stretch out on the hard floor of the cabin and sleep as soundly as an infant in the cradle! The motorboat trip was followed by an equally strenuous spell in a heavy open boat that had to be rowed a mile across the bay which at this point was too shallow for the motorboat. Reaching the shore of Epinette Point, we plunged into a dense forest of spruce and fir. For an hour we trudged and struggled in the rain through the meanest kind of tangled, boggy woods till we came to the more open part of the swamp. Examination of the ground led to the conclusion that bulldozers could clear away the swampy top soil and lay bare a foundation of sorts for landing strips, but only in case nothing better could be found.

Captain Roosevelt had warned us to keep together in so dense a forest; as we started back toward the shore he led the way, striding vigorously through the tangled underbrush, and as I brought up the rear it took some exertion to keep him in sight. I soon noticed that his course kept curving to the left—perhaps because of some obscure hereditary tendency. At frequent intervals I hailed him and asked for a compass check on our course to avoid traveling in a circle. When at last we reached the shore we were only 100 yards or so from where we had left the rowboat.

Next day we planned to visit a great elevated shoreline terrace on the south shore of Goose Bay, in the southwestern corner of Lake Melville, which I had photographed from the air just two years before. This seemed to be by far the most promising site in the entire region and had already been chosen by the Canadians for intensive study. But when morning came, the wind had backed into the north and was blowing hard. The skipper of the Hudson's Bay Company motorboat pronounced the going too rough for such an undertaking. By afternoon the weather was still rugged and Captain Roosevelt, who had set out in another boat for Goose Bay, was forced to return and await the morrow.

The following day, July 4th, the weather was clearer, but still the wind blew hard from the northwest. At last we found a man with a motorboat capable of taking an exploring party to Goose Bay. The round trip, including walking over the ground, was estimated to be a matter of nine or ten hours. The radio brought us a better weather report for flying, and while Captain Roosevelt and his party departed on the day's expedition, I was busily occupied superintending the removal of a wounded trapper from Northwest River down to Cartwright where he could get the surgical aid so urgently needed—aid that we were unable to administer with our own limited facilities. This trip proved to be more exciting than we had bargained for. The air, which was bumpy at best, grew worse and worse as we approached Cartwright. We neared our destination in the midst of a violent gale and rainstorm. One unusually bad bump threw the poor trapper right up from his bunk nearly to the roof of the plane. I was tossed off my feet and my camera was thrown around like a football. It was a relief to see Dr. Forsyth waiting for us on the wharf after a difficult landing. The trapper was quickly cared for and I subsequently heard of his complete recovery.

Although we were all inclined to stop at Cartwright for the night, Lieutenant Camera, in charge of the plane, received radio instructions to come on through to Newfoundland. Bidding the Forsyths and the trapper farewell, we were soon up in the clouds, and with the forbidding terrain out of sight below us, we flew on instruments. Presently the wings began to ice up and we came down a bit; patches of land and water began to appear below us and then we came right out of the clouds and saw the whole of the jagged shoreline stretching off into bright, clear weather ahead.

Many familiar landmarks of 1931 came into view and in the Strait of Belle Isle the greatest display of icebergs I ever saw. We emerged into bright sunshine and had a glorious ride back to Gander, where they put the photographer and me ashore with all our gear and took off again just at sunset for Argentia.

Captain Roosevelt and the others arrived the next day in the flying boat. They had returned from Goose Bay to Northwest River after midnight, weary but well satisfied that they had found on the great terrace much the best airport site in all that region. All next day and far into the night Captain Roosevelt worked on his report of the trip. A darkroom mounted on a trailer arrived, and the photographs were speedily developed, so that they could accompany the report, and on July 7th this report was dispatched by plane to Washington.

The Canadians, as we have seen, were surveying Goose Bay before we arrived. Captain Roosevelt, after studying the problem, became convinced that they were not prepared with equipment and manpower to develop an airport soon enough or on an adequate scale to meet the emergency. He therefore proceeded on the assumption that the task would fall on the U. S. Army Air Corps.

2 Chimo and the Torngat Mountains

We lost no time at Gander preparing for the next trip north to examine the region of Fort Chimo on the Koksoak River, just south of its estuary on Ungava Bay in far northern Quebec, and the mountainous terrain of northern Labrador. The choice between these two regions would probably depend on our observations, and we hoped that while there we could actually pick a good airport site in one of them. In either region an airport would provide the necessary link between newly-chosen Goose Bay and Baffin Island, which was to be the site of the next base in the northern air route.

Although the flying boats were back from Argentia on July 7, head winds, fronts, low clouds and rain kept postponing our departure. At last on July 12 we were roused at 3:10 A.M. to prepare for an early start. This time there were five Naval officers in charge of the flying boats, bringing the entire personnel of the expedition to twenty. I was assigned to the leading boat with Captain Roosevelt, to help in identifying places in Labrador.

Leaving the surface of the lake, we climbed through great masses of clouds till at 13,000 feet we reached clear air; the chill that penetrated the aluminum hull of the plane made us put on all the winter clothes we could lay hands on, and still we were cold. But once past the cloud belt we came down to where it was warmer.

Passing over Rigolet at 9,000 feet, we had a glorious view of Lake Melville and of Double Mer, stretching many miles to the westward in a blue band of exquisite color. How easy to see its true position from the air and correct an erring map! At sea in the northeast were vast quantities of pack ice, apparently choking completely the approaches to Hopedale, which on the same date in 1931 we had found wholly free of ice. Leaving Lake Melville far behind, we flew for hours over the vast interior of the Labrador peninsula, crossing hundreds of lakes and rocky ridges, the trees, at first dense, becoming sparser as we went north and neared the shores of Ungava Bay.

None of us had ever seen Fort Chimo or had been anywhere near this region. Our maps were so crude that they gave little help except to identify a major feature such as the George River, which we crossed obliquely, steering northwest. As we contemplated the enormous territory spread out before us we wondered a little whether we should find so small a spot as Fort Chimo. Soon another large river appeared and we saw its wide estuary

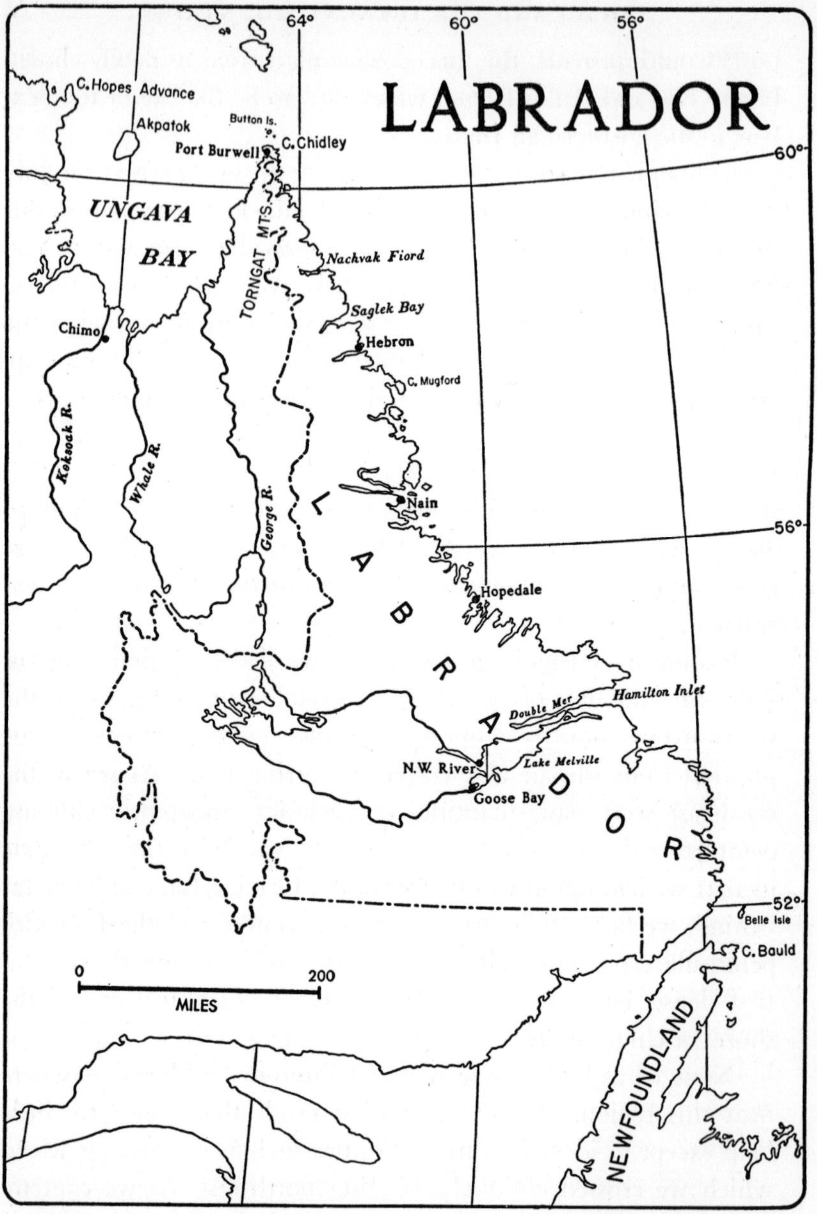
LABRADOR
64°
60°
56°
60°
56°
52°
C. Hopes Advance
Akpatok
Button Is.
C. Chidley
Port Burwell
UNGAVA
BAY
TORNGAT MTS.
Nachvak Fiord
Saglek Bay
Chimo
Hebron
C. Mugford
Koksoak R.
Whale R.
George R.
Nain
L A B R A D O R
Hopedale
Hamilton Inlet
Double Mer
Lake Melville
N.W. River
Goose Bay
Belle Isle
C. Bauld
0
200
MILES
NEWFOUNDLAND

to the north; we turned and flew north until we could clearly see the blue waters of Ungava Bay and had made sure the river was not the Koksoak. Turning west, we came at last to a great river flowing north into the bay; surely this must be the Koksoak. We turned again and followed it south till the last of our doubts were dispelled. With a thrill we saw the cluster of white, red-roofed houses of Fort Chimo, the only white man's residence in this whole vast area. For a few minutes we circled over the river banks, taking note of all areas that might serve our purpose, and then set down on the wide river just in front of the Hudson's Bay Company post. We had been seven and a half hours in the air.

Fort Chimo has long been one of the major trading posts of northern Canada. A hundred years ago it was more populous than today, but still the neat, white houses suggest a substantial unit of civilization in contrast with the complete wilderness extending in every direction. The Koksoak is one of the great rivers of the continent, bigger than the Hudson or the Delaware, nearly the equal of the Ohio; at Chimo it is a mile wide, offering anchorage for ships of considerable tonnage. Several auxiliary sloops, 30 feet or so in length, operated by both white men and Eskimos, ply the waters of the river, transporting whole families to the trapping and fishing grounds. A hundred or more Eskimos, with Chimo as a base, do a substantial business of trapping. Much of the year they live in tents or small huts at the settlement; the rest of the time they scatter farther afield and live a nomadic life.

As Mr. Edmunds, the interpreter for the Hudson's Bay Company, ferried us ashore in an auxiliary sloop, we saw a large throng of Eskimo women in brightly colored shawls watching us from the river bank above the rough log wharf. As we came ashore, each in turn came forward and shook hands cordially.

The Hudson's Bay Company agent customarily invites such rare visitors as find this remote spot to partake of his hospitality; and so it was with us. But we were a large party for his resources, and therefore Mr. Wenham, the resident Anglican clergyman, and his wife came over from their home near by to help prepare a dinner commensurate with our appetites.

At midnight we went to the abandoned storehouse of Revillon Frères and watched the assembled Eskimos of the settlement hold their Saturday night dance. The Eskimo lady who played the accordion provided rhythm and little else, but the square dances were well organized and executed. Mr. Wenham looked on, and when I commented on the contrast between his attitude and that of some of the Moravian Brethren on the Labrador coast where dancing had to be clandestine, he replied, "I've looked all through the Scriptures and found nothing in them against dancing."

All hands slept late next morning, for the tide would not serve our quest till noon. At Chimo there is a tidal range of nearly 20 feet, and when the tide is low, the wharf is an expectant gesture with about 200 yards of boulder-strewn flats (between it and the water) and on this the boats lie careened, awaiting the return of the tide. Man is a slave to the tidal cycle and our trip up the river had to be timed accordingly.

After attending services in Mr. Wenham's little church, we embarked in Mr. Edmunds's sloop, accompanied by the clergyman and Mr. Calder, the Hudson's Bay agent, and on a rising tide went 5 miles up the river past steep-sided islands. The sloop anchored off a long beach of brown sand backed by a steep bluff on the west side of the river. A fine flat area of dry, sandy soil with caribou moss and scattered dwarf birches stretched back half a mile from the bluff. Crossing this, we climbed a small rocky knoll on the south side, from which we studied and photographed the terrain. Captain Roosevelt then assigned each of us an area to examine for runway possibilities, with instructions to pace it off and report the length of a possible landing strip. Mr. Wenham, eager to help, did his full share of the exploration. That evening we drew sketch-maps showing roughly how three runways of at least 5,000 feet each could be fitted into the area west of the river.

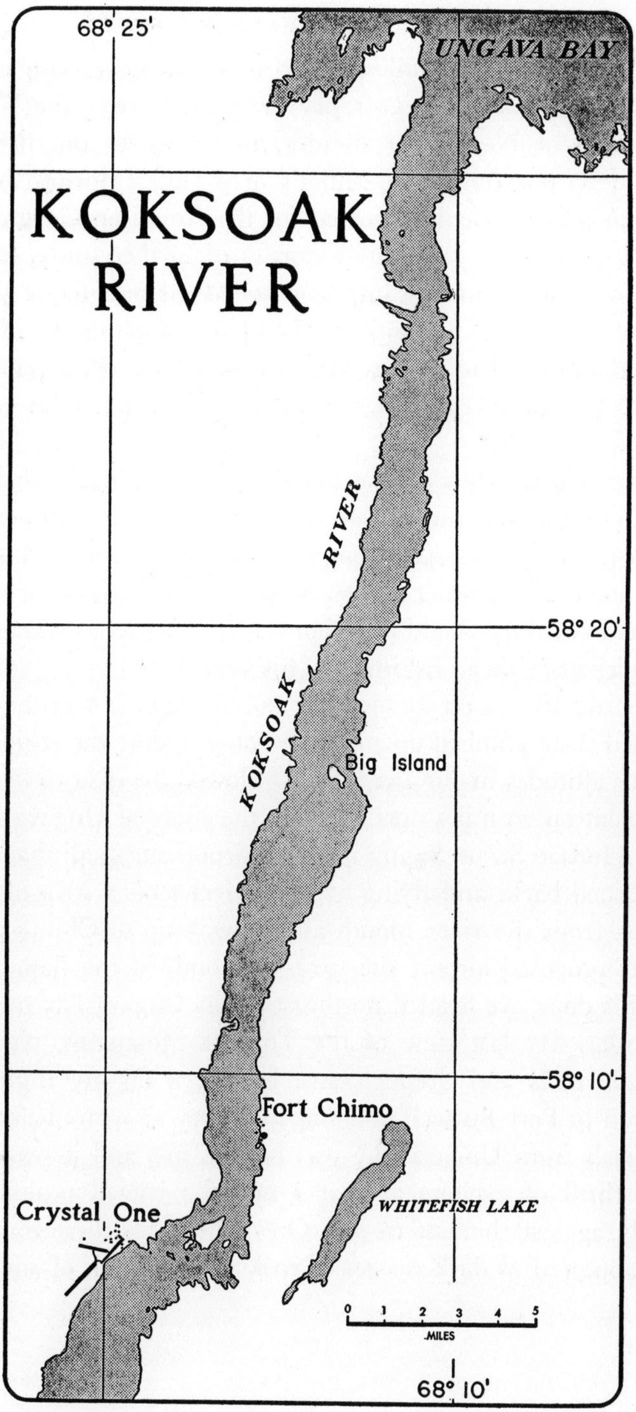
68° 25'
UNGAVA BAY
KOKSOAK
RIVER
RIVER
KOKSOAK
Big Island
58° 20'
58° 10'
Fort Chimo
Crystal One
WHITEFISH LAKE
0 1 2 3 4 5
MILES
68° 10'

July 14 dawned cloudless and continued so over Ungava Bay and most of Labrador. Our plan was to fly over the Torngat Mountains of Northern Labrador, to look over the flat areas marked on our tracing of Miller's map (made during the 1931 research project I had organized for the American Geographical Society), and if any appeared worthy of further study, to come down on a fiord, make camp, and spend the next day exploring on foot. But first it was important to photograph the Chimo area from the air and to look well at the shores of the river all the way to the sea, in case other areas might be better than our first choice.

The tide was low in the morning and our packs had to be carried far out over the wet sands to a rowboat in order to embark. To hasten the process, six strong Eskimos carried a heavy but leaky dory down to the water's edge. Most of us were shod for wading to where the dory would float, but Captain Roosevelt was not, and rode pickaback on an Eskimo half his size.

Shortly after noon we took off, and zigzagging over the island-studded river, climbed up and up, photographing the region from various altitudes in our ascent. We followed the river to its mouth and went on 20 miles northwest up the shore of Ungava Bay toward Hudson Strait. Seeing no good airport site in all that region, we turned back, and flying at 10,000 feet took a strip of photographs from the river mouth all the way up to Chimo and on to our proposed airport site, 30 miles south of the Bay.

This done, we headed northeast across Ungava Bay to look at Labrador. My last view of the Torngat Mountains, with their gigantic cliffs and deep chasms, had been on my flight from Hebron to Port Burwell and Cape Chidley six years before. The approach from Ungava Bay was new to me, and it was with a keen thrill of excitement that I saw the jagged peaks appear faintly against the eastern sky. On earlier trips these mountains had appeared as dark masses of rock with a patch of snow here and there. Today they were clothed with almost unbroken snow, and the picture they presented was magnificent.

Such is the illusion of height when a sharp mountain profile rises above a distant horizon that, although Miller's careful triangulation showed no peaks in all northern Labrador to rise much over 5,000 feet above sea level, Captain Roosevelt and the chief pilot could scarcely be convinced that the peaks before us were not near to our present altitude of twice that figure. With nearly full throttle the plane was gradually coaxed up to 11,000 feet to insure a safe crossing. Indeed so apprehensive was the pilot that he warned me not to move about in the navigator's compartment lest I disturb the equilibrium of the plane in its struggle to clear the mountains—and this in a plane weighing many tons! Actually our altitude was more than 6,000 feet above the highest peaks in our path.

Presently the peculiar truncated outlines of the peaks in the central range could be made out, and beyond these lay the large flat area that photographs had indicated as the best prospect hereabouts. The peaks around Komaktorvik Lakes were identified and beyond them Mt. Tetragona lying between two converging fiords. Now we were over the broad valley connecting the heads of these two fiords, hemmed in between Tetragona and the central range. We were more than 10,000 feet above the valley floor, but the pilot, accustomed to southern waters and gently sloping shores, shuddered at the sight of frowning cliffs and snowy peaks, with narrow fiords reaching out to an ice-choked sea. Though the heads of the fiords offered wide reaches of open water, he would not dream of coming down to land on them, nor would he sacrifice any of his hard-won altitude for a closer view. Captain Roosevelt scrutinized the ground below as well as he could and concluded that although an airport could be placed in this valley, high mountains with precipitous flanks were too close for safety, and he would not recommend it. He therefore acquiesced in the pilot's view that we should head south without further delay, and with the other plane following behind, we started our long journey back to Newfoundland, where we stopped at Botwood for the night. Here we learned that Le Breton's task force had occupied Iceland.

Early the next morning we landed on the lake at Gander and Captain Roosevelt, without even stopping for breakfast, went right to work on his report of our reconnaissance with its recommendation of the Chimo area. All day and nearly the entire night he was busy, while the photographers developed their pictures and Mr. Preston worked on his drawings. As soon as it was finished the report was to be flown to Washington. Captain Roosevelt was to accompany it, and I was to go as far as Boston and there report back to the First Naval District Headquarters. The plane was scheduled to start at 2.30 P.M., a bare 30 hours after our arrival from the far north.

It was not long after the appointed hour that we were loaded into an Army bomber and taxied out to the great runway. Five Hudson Lockheeds were lining up to take off on their nonstop flight to England. We waited and watched them get off one by one and disappear into the east, to begin their combat careers, and then took the air ourselves. A front of dense clouds over Nova Scotia diverted our course to the northward and we passed over the Magdalen Islands. One of these, close beneath us, was a wonderful sight with its emerald green rolling hills, dotted with white farmhouses, and its superb red cliffs, set in a deep blue sea, all gleaming in the brilliant afternoon sun. Soon we came to Campobello, where Captain Roosevelt extended to his grandmother a cordial greeting by getting the pilot to "buzz" her house at about 300 feet. The view of the Roosevelt domain there was detailed but brief. Familiar islands from Machias to Casco Bay passed in quick review and by twilight we had landed at the Boston airport. There I bade farewell to Captain Roosevelt, who went on to Washington with the report.

At the War Department the receipt of the report was followed fairly promptly by the decision to adopt the site we had chosen on the west shore of the Koksoak River above Chimo for a prospective airport.

3 The Trawler Fleet; Halifax

No further word of the Northern Air Route came to me until September 19, 1941, when I heard that a fleet of trawlers was to leave Boston two days later to meet an Army Transport at Halifax and proceed north with personnel, radio equipment, portable houses, and a winter's supply of food and fuel for the advance bases that were to be established at Fort Chimo and Baffin Island, the next point along the route. Captain Paul Grening of the Army Transport Command was frantically trying to locate pilots for the fleet, for the Water Transport Service of the Army had efficiently assembled every conceivable item needed for an Arctic trip, but had not managed to unearth a local pilot. Names of qualified persons that I had furnished in August were already buried in the depths of Army files and there were only thirty hours left until departure time! Captain Grening moved heaven and earth to get me released from my assignment at the Squantum Base. I was eager to go, for the lure of the North was still strong and I had waited two months for just such a piece of luck. After a hurried trip to Washington filled with hectic conferences, interminable questionnaires, and the usual amount of red tape I was finally sworn in as an Arctic Pilot in the Civil Service and sent on my way rejoicing. "Shouldn't I have orders?" "No, go right ahead; they'll be sent after you." Although the trawlers had departed on schedule September 21, I was able to get a train to Halifax and met the small fleet there.

Halifax was one of the great nerve centers of the war whence large volumes of vital shipping moved eastward; just to be there was a stirring experience. The days that followed were crammed with a turmoil of busy preparations and excitement as the time of our departure on this new adventure drew nearer. The Furness-Withy wharf was alive with activity, for there was much loading to be done and it might be a matter of several days before all the final arrangements were completed.

My old friend Charles J. Hubbard with whom I had made flights in Labrador during the mid-1930's had recently been commissioned Lieutenant Commander in the Naval Reserve. He had been actively organizing the expedition in New York at top speed for three weeks and had just arrived in Halifax aboard the Army Transport *Sicilien.* She carried a full load of winter supplies for the three different bases, each to be garrisoned by ten men. Radio equipment for constant communication with Washington, and instruments for weather records were the two main items. Airplane fuel was included for such planes as could land on the ice in winter, but no actual airport construction could be started before the following year. Colonel Wimsatt of the Air Corps was the military commander of the expedition, while Captain Grening (the person most responsible for my being present at this undertaking) represented the Quartermaster Corps of the Army in a civilian status, with authority to make all arrangements for water transportation.

Seven Diesel-powered trawlers had assembled here in Halifax: five, the *Fabia,* the *Flow,* the *Cambridge,* the *Lark* and the *Cormorant,* were of the Boston fishing fleet and had come from there under Grening's command; two others, the *Polarbjorn* (Polar Bear) and the *Quest,* were Norwegian ships belonging to Mr. Carlson, a Norwegian then in Halifax. All were to take on most of their loads from the *Sicilien* here in Halifax and then follow her north to a final distributing point where they would scatter in three groups,—Number 1 to Chimo, Number 2 to Frobisher Bay, at the southeastern corner of Baffin Island, and Number 3 to a point on the Northeast coast of Cumberland Peninsula, well up the Baffin Island shore, some 230 miles nearer Greenland than the base in Frobisher Bay.

The orders from Washington directed the *Sicilien* to proceed with the whole trawler fleet to Hebron, far to the south of Cape Chidley, and there to divide into the three groups and proceed separately to the three bases. Somehow Hebron had made an impression in Washington as the northernmost settlement on the Labrador coast, and the impression had stuck. Hubbard, with his intimate knowledge of Labrador, well knew that this choice was a bad one, for Cape Chidley with its formidable tide rips still lay 160 miles ahead beyond a dangerous coast beset with deadly reefs. At the time of separation each trawler must take on a heavy deck load from the *Sicilien,* and, worse still, one ship of each group must take a tank-lighter in tow for the last lap of the journey. Each tank-lighter carried a heavy tractor to be put ashore at the base. The men who were to occupy the bases through the winter and the men of the Engineer Corps who were to erect the houses must be transferred from the *Sicilien* and live aboard the already crowded trawlers. Clearly the separation point should be as near the bases as possible. Port Burwell, at the northeast corner of Ungava Bay, around the point from Cape Chidley, where Hubbard and I had stopped for lunch on our flight in 1935, was the obvious place for rendezvous and final division of the fleet. Hubbard urged this point with earnest emphasis. Colonel Wimsatt, a competent Air Corps Officer, but with little knowledge of the sea or ships and no firsthand knowledge of Labrador, recognized the value of Hubbard's seafaring and local knowledge, but he hesitated to demand a change of orders from Washington. Throughout the first day this question hung fire.

The unloading of the *Sicilien* was a sight to see, as well as an engineering feat of no little skill. The dock presented a scene of tumult such as I had rarely seen, as great crates of winter supplies and equipment were swung ashore and quickly sorted according to destination. Before leaving New York every crate had been marked with the designation "Crystal" and numbered 1, 2, or 3, showing to which base it belonged. The commanding officer of each base stood by to see that his gear was correctly segregated. Crystal One, by Fort Chimo, was to be in charge of Lieutenant Commander Schlossbach, a man of much winter experience in polar regions and for a time associated with Sir Hubert Wilkins. Crystal Two, in Frobisher Bay, was under Major John Crowell, who had sailed as mate of the schooner *Thebaud* with Commander Donald MacMillan on an exploration of that bay some years ago. Crystal Three, the farthest north, was to be in charge of Captain Dyer of the Army. I found Edward Goodale, who had served under Admiral Byrd as sled-dog expert in Antarctica before his cruise to Labrador, appropriately attached to the northernmost group and busy with the others at the sorting of cargos. To add to the din of unloading, there was on the dock a whole village of crates containing sled dogs for winter transport, howling like a thousand demons.

By September 25 the plans were fast taking shape and the loads removed from the *Sicilien* were being stowed aboard the trawlers. Hubbard, fearing that the seven ships on hand would not suffice to carry the load, started a search for another, and soon found that Carlson, owner of the *Polarbjorn* and the *Quest,* knew of a similar ship, the *Selis,* belonging to the Norwegian Navy and now at Lunenburg, an eight hour's journey to the west along the coast. Colonel Wimsatt decided to take Hubbard's advice on two important points and sent coded radio requests to Washington for authority to charter the *Selis* and to keep the fleet together all the way to Port Burwell before dispersing to the three bases. The *Sicilien* would accompany group Number 3, going farthest north, so that from Port Burwell the towing need be done only across

Ungava Bay, in the case of Number 1, and from a second distributing point in Frobisher Bay, in the case of Number 2.

It was decided that I should start ahead with the first five trawlers to be loaded, including those of slowest speed, and proceed to Curling, Newfoundland, for refueling. We should be overtaken by the faster ships either there or at Hebron. This was confirmed by orders from Colonel Wimsatt.

Later, with the prospect of delay in loading the *Polarbjorn* and the *Quest,* and of further delay in getting the *Selis,* my fleet was reduced to the *Fabia,* the *Flow,* and the *Cambridge,* the three bound across Ungava Bay to Fort Chimo for Crystal One. On my flagship, the *Fabia,* the skipper was Johannes Bjornson, a Danish seafaring man, born in Iceland, but now an American citizen and veteran of many a hard winter's fishing cruise on the Banks. The mate spoke in warm praise of him; all signs seemed to show that the *Fabia* was a "happy ship," and so she proved to be.

My task during the loading period was to look over the charts, sextants, and other navigating gear on every ship and see that all were properly equipped; also to investigate anchor chains and see that each trawler had enough. Trawlers in their fishing routine rarely anchor; they travel and drift far out at sea, and returning with full holds, tie up alongside the fish pier. Thus, ground tackle is to them a sort of insurance against the unusual. But Grening well knew how vital stout anchors and chains with ample scope would be in the windy harbors of the north, and he left no stone unturned to be assured of enough ground tackle for security. One example of his forethought was the suggestion that on some boats the binoculars should not be put aboard till the last minute, lest they be smuggled ashore and sold, and the proceeds transformed into rum!

An earnest request came to Grening from one of the Army garrisons for victrola records of Beethoven's Fifth Symphony to make life livable in the lonely camp during the long winter. They already had some sonatas, but they longed for that symphony as well. He told me the only way he could get the records out of the Army appropriations was to call them shovels—a far cry from "calling a spade a spade."

The next day, oath-provoking messages came from Washington, including one suggesting that the Coast Guard Cutter, *Comanche,* be sent from Greenland to join us at Hebron, and that we should take 500 barrels of oil from Halifax to refuel her when we should meet. This amount of oil was not available, and even if it had been, the delay for refueling at Hebron was one we could ill afford so late in the season. Thus the *Comanche* would be a liability rather than an asset, and Washington was so advised. Still no reply came to the urgent requests about the *Selis* and Port Burwell. Finally, Colonel Wimsatt in desperation telephoned to Washington, and in trenchant terms called for action. Before he hung up he had a green light on both points. We left the office on the run in search of Carlson and soon found him. He sent a dispatch to the *Selis* to come at once to Halifax. She should arrive by noon next day, Saturday, and be loaded and ready to start early Sunday morning.

Progress in loading had seemed to indicate a probable departure for the first group, in which I was to sail, before noon next day. To my surprise at 5:30 P.M. Colonel Wimsatt said to me, "How soon can you get going? The *Flow* will be loaded at six o'clock and ready to sail." A mad scramble to get started ensued. The *Flow* WAS loaded by six o'clock, but no one had told the crew to stand by, and they had scattered all over town. Also we had to be cleared for departure by the Naval Control Service, and a pilot must be put aboard. None of this could be done in time for departure before daylight next morning. But our sailing was definitely scheduled to follow close on the dawn.

Fort Chimo on the Koksoak River, looking upstream

Cape Chidley, October 1941

The Sicilien *and the trawler fleet in Port Burwell, October 1941*

Norwegian trawler, Port Burwell, October 1941

Killinek Eskimos, Port Burwell, October 1941

The site of Crystal One, Koksoak River, October 1941

Tank lighter with tractor and lumber makes first landing at Crystal One, October 1941

Eskimos watch the landing of the tank lighter

4 Northward to Ungava Bay

In mid-July, on our flight back from the Torngat Mountains, we had seen the sea all along the Labrador coast completely choked with pack ice. Inquiry of the Hudson's Bay Company officials now brought the good news that this ice was all gone and the sea clear for navigation. The pilot was scheduled to report on September 27 at 6:00 A.M. to take us out of the harbor. The sailing of three trawlers out of Halifax was no unusual event, but to me the command of such a fleet, bound for the far north on such a mission was so novel and exciting an assignment that I was awake and up long before dawn. Not a soul was stirring. About six o'clock, as a faint streak of dawn was appearing in the east, Hubbard and Grening came again to the dock to say goodbye. The crews of the trawlers turned out, the pilot came at last, and at 6:45 A.M. we backed off from the wharf in the *Fabia* and started down the harbor, followed by the *Flow* and the *Cambridge*.

Soon we had dropped the pilot, passed Chebucto Head, and were headed east over the heaving swells of the broad Atlantic. On the *Fabia* we set the course; the *Flow* and the *Cambridge* took station 500 yards away on the port and starboard quarters and followed our lead.

The *Fabia* was a staunch, well-found wooden ship, better for coping with ice than the steel ships that accompanied us. The crew of eleven included the captain, mate, chief engineer and two assistants, a radioman, cook, boatswain, and three seamen. The galley was a cheery place where the cook placed great dishes of good food on the long table and all hands helped themselves. They were a well-fed and happy crew, and it was good to be with them. All but one were born in lands ranging from Nova Scotia to Iceland and Italy, but all were now United States citizens with homes in and about Boston. A young Army radioman, "Red" Newell, had come abroad from the *Sicilien* to relieve the ship's radioman, so that between them they could stand continuous watch. Newell was scheduled to winter in the north.

Doing an honest 9 knots, we passed Jeddore Rock, Liscomb Light, and other familiar landmarks during the day, and by midnight were well past Cape Canso.

The next day was so clear that soon after we had dropped the high land of Cape Breton Island over the western horizon, Newfoundland appeared on the starboard bow. Before ten o'clock that night we had brought Red Island light abeam, and slowed down to 7 knots in order to reach the Bay of Islands in the dawn. The high cliffs of the entrance loomed up very black and impressive against the streaky gray dawn. So bold were the shores, there was no difficulty in seeing them, even in this dim light. Going "dead slow," the ship rolled heavily till we entered the bay and smoother water. As we neared Curling we were stopped and boarded by a pilot, who said his presence was compulsory, though there was nothing for him to do. It was full daylight when we tied up to the Imperial Oil Company dock at Curling, where the ships were refueled for the long voyage northward.

One of our crew was perturbed at seeing a rat board the ship from the oil dock; but the Skipper said cheerily, "Oh, that's our rat. He always goes ashore at every port to see the sights."

At the dock a wire was handed me from Colonel Wimsatt instructing us to remain here after fueling, pending further instructions. This canceled the previous orders under which we might have proceeded to Hebron before meeting the others.

We awoke October 1 to find the *Sicilien* anchored out in the bay and the rest of the trawlers tied up at an adjoining dock. The *Selis* had come from Lunenburg, loaded at Halifax, and made the journey with them, but her status was still unsettled, for she belonged to the Norwegian Navy and her release had not yet been fully authorized. A radio telephone call to London solved the problem and the *Selis* became a full-fledged member of our expedition. That evening after dark the trawler fleet moved down Humber Arm to an anchorage by Wood's Island.

The morning of October 2 dawned brightly with the clouds breaking away and patches of pink sunlight touching the hills. When the *Sicilien* gave the signal for departure, the *Fabia's* groaning windlass slowly heaved up the anchor, bringing with it another anchor, ancient and vast—big enough for a cruiser. It took us half an hour to get rid of this incubus, and then full speed was needed to overtake the procession already well on its way to the Strait of Belle Isle. Like a brood of ducklings following the mother bird, this strange assortment of trawlers followed in the wake of the *Sicilien*. But we were not a very orderly brood, for the Norwegian ships were unable to maintain the speed of the larger trawlers. Grening had reckoned with this, and sailing aboard the *Lark*, he assumed the role of chaperone to the stragglers, staying with them and holding them to the course when we had disappeared over the horizon ahead. As I learned in later years, Captain Grening could quickly get lost within 50 yards of his office in the Pentagon Building in Washington (who couldn't?), but he knew how to find his way at sea.

The first four days of our voyage northward were uneventful. It was with pleasant memories that I watched the old, familiar landmarks pass in review. In turn, as we sailed along the coast of Labrador, Mt. Thoresby, the Kiglapaits, and the Kaumajet Mountains of Cape Mugford were recognizable, and on the morning of October 6 the sun was shining full on the eastern face of the Torngats. Straining my eyes eagerly through powerful binoculars,

I identified peak after peak that had towered above our schooner as we had groped for the channel on that memorable voyage of 1931.

Snow squalls were now moving fast over the face of the mountains, and it was well to know our position, in case the squalls should envelop the mountains—and us. They soon did. Using Miller's map and my sextant I was able to get a positive fix of our position. For a brief time the snow squalls ahead parted and revealed the stupendous cliff of Cape Chidley, now some 25 miles ahead. As if aware of its importance to us as a turning point, its snow-covered crown gleamed brightly as a beacon in the sunshine.

Before noon the snowstorm struck in earnest, reducing visibility to scarcely more than a quarter of a mile. The *Sicilien* slowed almost to a standstill, blew her foghorn, radioed us to keep in sight of her, then turned and headed out to sea. Her skipper didn't fancy closing with Cape Chidley in such a storm. In an hour or so the squall passed and Cape Chidley reappeared; we resumed our northwesterly course, and though more snow hid the mountain summits, the Cape remained in view, a truly marvelous sight as we approached. We passed the great cliff and entered Gray Strait just as the tide was turning to run west and help us through. On our port hand the giant headlands of Killinek Island were cloaked in formidable storm clouds; to the north the Knight and Button Islands, covered with fresh snow, stood out clear and bright in the afternoon sun.

Captain Barbour from St. John's had come to help us with the pilotage, for he had been on sealing voyages in these waters. He had warned us of a reef off Flat Island, west of Cape Chidley. We needed no warning; a heavy sea was rolling into the strait from Ungava Bay, the tide rip off Flat Island was tremendous, and the sight of the tumbling breakers all around the reef was enough to warn any skipper to give it a wide berth. In the deepening twilight we of the vanguard rounded the southwest corner of

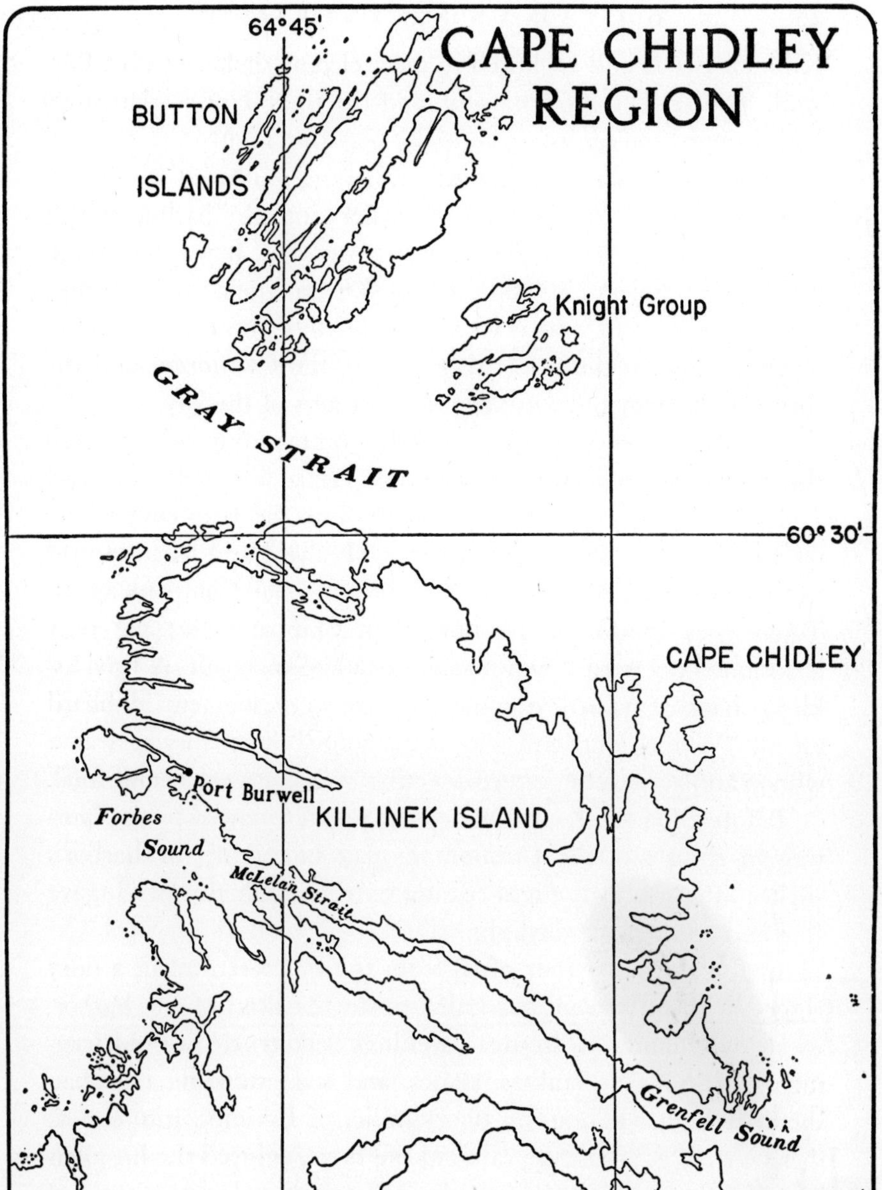
CAPE CHIDLEY REGION
64°45'
BUTTON ISLANDS
Knight Group
GRAY STRAIT
60° 30'
CAPE CHIDLEY
Port Burwell
Forbes Sound
KILLINEK ISLAND
McLelan Strait
Grenfell Sound
60° 15'
0
5
10
Miles
64° 30'

Killinek Island and anchored in the welcome shelter of Port Burwell. The *Lark,* the *Cormorant,* and the three Norweigian ships had dropped too far astern to make port that night.

October 7 was a grim day. An icy, strong southwest wind blew in from Ungava Bay sending big rollers into the harbor, which offered but imperfect shelter from that quarter. It was too rough to lay the trawlers alongside the *Sicilien* for their final loading; we just had to wait, and as we waited our thoughts turned to Captain Grening on the *Lark,* shepherding the *Cormorant* and the three Norwegian ships in the stormy waters of the bay.

On the *Sicilien* were quartered the commanding officers of all three of the prospective bases. From her a dory put out with Hubbard and Wimsatt aboard, rowing into the snug cove where on a calm sunny day in August 1935 Hubbard and I had landed in the seaplane for lunch with the Hudson's Bay Company agent. Today they found the Hudson's Bay Company post not only abandoned but torn down and not a white man left. A tiny Eskimo shack remained, occupied by two aged women. Hubbard asked, "Where Hudson's Bay men gone?" The answer was a whoop and a spacious gesture denoting exodus complete and final.

Before noon the *Lark* rounded the point, followed by the Norwegian ships and the *Cormorant,* glad indeed of the harbor's shelter after a beastly night tossing on the rough seas of Ungava Bay and waiting for daylight.

In the afternoon four of us from the *Fabia* set out in a dory rigged with a spritsail and sailed up to the head of the harbor, where we found two strange dwellings (constructed of old fragments of lumber, blankets, stones, and sod) standing less than shoulder high, yet housing two families of Eskimos. Into one of these we crept, and sitting around the stove, enjoyed the hospitality of these ever-friendly people. No common language served to promote sociability, but through the medium of tobacco and sharing the warmth of the stove and the greater warmth of the humanity which radiated in the genial smiles of our hosts, there was companionship enough. Newell and I climbed the steep snowy slope to the crest of the ridge whence we could see to the north, beside a snug inner harbor, the abandoned Moravian church with its tall steeple and to the south the trawler fleet at anchor clustered around their mother ship. "Port Burwell" is now a strange-sounding name for this desolate harbor of refuge, without docks, buildings, or anything else to greet the eye save the two or three Eskimo shacks, barely distinguishable from the surrounding rocks peeping from beneath their mantle of snow.

That afternoon the wind and sea went down and before dark the *Fabia* received a signal to come alongside the *Sicilien* and take on her final load for the journey to the Crystal One base at Fort Chimo. A tremendous load of portable house sections was piled high on the *Fabia's* deck, and her tanks filled with fuel oil. I climbed the rope ladder to the *Sicilien's* deck and conferred with Colonel Wimsatt about orders for our forthcoming journey. Before drawing them up for typing, he showed me a dispatch he had received from Washington telling him that my questionnaire revealed nothing about my qualifications as Arctic Pilot, and asking what they were. The questionnaire had given me no opportunity to offer enlightenment on this one essential point; just how they expected a revelation remains a mystery. Wimsatt replied with a reference to the survey of 1931.

Another message came asking if we were meeting the *Comanche* at Hebron (now about 200 miles behind us). Wimsatt snorted and sent back a caustic reply. I remarked that the *Comanche* seemed to be the standing joke of the expedition. "It's a damn poor joke," said the Colonel, "I fail to see the humor of it."

5 The Koksoak River; Crystal One

Loading the three trawlers, *Fabia, Flow* and *Cambridge,* bound across Ungava Bay for Fort Chimo, took till October 9 to complete. The tank-lighter with the tractor stowed inside was a heavy burden for the cargo boom, and as it swung over the side, the *Sicilien,* big as she was, listed sharply to starboard. We were to tow this clumsy craft astern of the *Fabia,* and to protect the tractor and prevent swamping, a large tarpaulin had to be nailed down, enveloping lighter and cargo.

While this was going on, it was decided that Hubbard should take charge of the *Cormorant* and proceed north in her ahead of the fleet in search of the site for Crystal Three, the most northerly of all our bases. It was an adventurous project, for the journey was long, the east coast of Baffin Island was wild, uncharted, and almost unknown, and the season was so late that one veteran of the north had already said, "If anyone considers going into those waters at this time of year, he had better forget it." But Hubbard went. First he had to find his way to Pangnirtung in a deep fiord far up Cumberland Sound, where lies one of the main settlements of the Canadian Arctic, for there lay the best hope of finding a local pilot. Pangnirtung, the home of many Eskimos, boasts a hospital with a doctor and nurses. Here Hubbard found a ravaging epidemic of influenza going on, and the one Eskimo who knew the east coast well enough to guide the ship to the chosen site was sick abed in the hospital and said to be on the verge of pneumonia. Here was a dilemma. Haste was urgent; a single day's delay might jeopardize the whole project. The *Sicilien* and the two trawlers assigned to Crystal Three might get caught in the fast-forming winter ice. But the base must be established. The doctor saw the point, and filling the patient with sulfa drugs, had him carried aboard the *Cormorant,* while the nurses cursed Hubbard for a villain. For his protection, Hubbard had a Canadian "Mounty" assigned to go with them as a witness to whatever might happen. They put to sea and in three days the Eskimo was up on the trawler's bridge guiding her to their remote destination. After picking the location and sounding the passage, Hubbard headed the *Cormorant* south again to meet the *Sicilien* as she started north after leaving the Crystal Two group of trawlers to find their way into Frobisher Bay. Under Hubbard's guidance, at dusk in a snowstorm, the *Sicilien* and the *Quest* proceeded to Crystal Three. Both of these units discharged their cargos, built the bases and brought back the Engineer Corps builders. And just in time, for by early November the waters were closed in the grip of winter.

As we on the *Fabia* watched the *Cormorant* head boldly out of the snug shelter of Port Burwell on that wintry October afternoon, we knew she was bound on a perilous voyage, alone and far from help. It is a great tribute to the daring of Hubbard and his band of brave men that they were able to accomplish their mission with speed and safety.

Late in the afternoon our own Crystal One Group was ready and in the gathering gloom we put to sea. As we passed the *Sicilien* she gave us a heartening salute from her deep-throated whistle, and in a few minutes she and all Port Burwell were lost in the murk astern. Plunging into the head seas kicked up by a southwest wind, the tank-lighter we were towing disappeared into a mass of spray with every wave. It was well that the tarpaulin was firmly secured!

Through the night we steered southwest, the *Fabia* in the lead, the *Flow* and the *Cambridge* following. Log readings were sent us by radio from the *Cambridge* every two hours for the purposes of dead reckoning. Commander Schlossbach, as officer-in-charge of Crystal One, was in command of our group. He and Captain Barbour, the pilot who had come from St. John's, were now on board the *Fabia,* still flagship of group Number 1. Barbour had once been up the Koksoak River to Chimo in a Hudson's Bay Company ship and thus had a smattering of local knowledge, but he said most emphatically that we must have a local pilot to take us up the river; to do without one would be to court disaster. My designation was "Navigator"; as such my duty would cease when we had crossed Ungava Bay and found the mouth of the Koksoak River.

The only chart of Ungava Bay then in existence was of the crudest variety. It suggested scattered reefs off the river mouth, but gave only vague clues as to even their approximate location. To dodge them by celestial navigation was highly problematical; yet any sights we could get of heavenly bodies should help us find the corner of Ungava Bay where the river was alleged to emerge.

Long before dawn I was roused, and finding the moon dimly visible through a thin place in the cloud cover, brought my sextant to bear and started making a series of precautionary observations. When the sun rose and came out into clear sky above a cloud bank, more sights were taken, and as I was plotting our position on the chart, land appeared ahead. Steering south, we sighted the beacon by the river mouth and entered at two o'clock. The tidal range here is about 30 feet, and the ebb tide, augmented by the river current, was pouring a tremendous torrent of water out into the bay. It was a problem to find a safe anchorage. Feeling our way with the fathometer, we worked in toward the shore out of the force of the current till Schlossbach was satisfied and gave the order to let go. But the fathometer showed only four fathoms and the tide had two more fathoms to fall. The ship drew 17 feet and some quick mental arithmetic convinced us that we would better get out. The anchor was weighed, not a moment too soon, for the ship bumped two or three times on boulders before she gained the deep water of midstream. The *Cambridge,* anchored close by, was not so lucky. She was hard aground with the tide falling fast!

We dropped anchor out in the current and for a time, by dint of watchful steering, kept the ship headed straight upstream, but presently the swift current caught our bow and sheared us out into the swiftest part, where we took it broadside—the anchor cable leading from the starboard side not much forward of amidships. This arrangement, tolerable in ships that normally keep the sea or lie at a wharf, is the worst imaginable in such a situation. The strain was too great for the brake on the cable drum, and fathom after fathom crept slowly out while the wooden brake bands started to smoke from the heat of friction. At last, by using the main engine and the rudder, the ship was got under control and the cable heaved in a bit. A spare cable was rigged through the hawse-pipe in the bow and the anchor secured to that. The ship then lay nose to the stream, as she should.

We watched the helpless *Cambridge* heel farther and farther on her side as the tide went down. About 7:30 the stricken ship, now in total darkness because she could no longer run her generator, sent a blinker message that she was taking water fast in the engine room. A second, more urgent message called for help. The tank-lighter was made ready and a dory launched, but one of the crew lost the painter and the dory started out to sea on the still ebbing tide. Fortunately the lighter was nearly ready and set out in pursuit. The dory was located with a flashlight, caught, taken in tow, and the lighter proceeded with her to the luckless *Cambridge.* Schlossbach stayed on aboard the stricken vessel reassuring her anxious skipper till midnight, when the rising tide floated her off.

We had managed to get a message relayed to Fort Chimo reporting our arrival and on the morning of October 11 Mr. Edmunds arrived in his boat with three Eskimos to assist him in piloting the trawlers up the river. Mr. Edmunds was the man who had furnished transportation and pilotage during our visit to Chimo in July, and he was a key figure in all of our operations on the Koksoak. Married to an Eskimo woman, he knew the language as well as he knew the river, and both funds of knowledge were of the greatest value in establishing the base at Crystal One.

On the powerful flood tide the three trawlers sped up the river, passing through whirlpools and by swirling eddies, as the readings of the fathometer varied abruptly between 10 and 40 fathoms. Edmunds and his Eskimo colleagues knew the tricky channel well, and so swift was the current that with the ships running their engines at moderate speed we traversed the 25 miles to Fort Chimo in two hours. There we came to anchor in front of the friendly cluster of red-roofed houses.

Next morning many sections of the portable houses were loaded into the tank-lighter in the space in front of the tractor, together with a big tent and food for the party that was to make camp at the new base. With Mr. Edmunds again serving as pilot, we started up the river on the flood tide to land the first load on the site we had chosen in July. We were accompanied by his eleven-year-old son, Freddie, smiling with the sunshine that lights up the friendly faces of the north. We passed the sheer cliffs of Elbow Island and made a landing on the steep, shelving rocks just beyond the prospective air field. A crowd of Eskimos greeted us and helped unload.

Day after day the transportation of cargo went on from the trawlers at Chimo to the base 5 miles up the river. The Hudson's Bay Company boat and several others belonging to the Eskimos were impressed into service. Twice I made the trip under sail in a sloop manned and inhabited by a whole Eskimo family who invited me to share the warmth of their tiny cuddy where a fire burned in a little stove. These good people turned to with a will and did a large share of the arduous work of unloading the cargos.

It was impressive to see a little Eskimo woman with a heavy baby slung in the hood of her parka, and a 30-pound box of canned goods in her arms, trudging patiently up the steep bank over and over again, carrying load after load. The extra work of carrying the baby up and down could not be avoided, for there was no safe place to park him where the dogs could not find and eat him!

For more than a week this work went on. Transporting large cargos up that river in small boats was slow business, and more than once the boats were hung up on shoals where they had to await the rising tide to carry on. Even at the base, 30 miles from the sea, the tidal range is about 12 feet.

In July I had photographed and mapped Whitefish Lake, whose size and proximity to Chimo commended it as a possible winter landing field. I now wanted to study the approach from Chimo to the lake, to see if a tractor could get there. Hubbard had suggested that empty oil drums filled with sand could be used to roll the snow on the frozen lake and thus make a runway fit for airplanes operating on wheels. A tractor needs no boulevard, and if the intervening terrain presented no formidable obstacles, the garrison could bring the tractor from the base and thus put a winter airport in operation.

Freddie Edmunds was assigned to me as a guide, and a better companion for a day's tramp I could not have asked. He carried a 22 rifle and handled it like a veteran. Firing at a well-concealed perch, he brought down a ptarmigan which had been invisible to me. He took me through a natural roadway to the lake, a gully with a level floor between rock walls, and but few obstacles to be surmounted. On the way in, scarcely half a mile from Chimo, I noticed and photographed a very large area of level swampy ground. Though it presented a surface inferior to the firm terrace at the base chosen in July, still it could be cleared, and its proximity to the anchorage at Chimo was a great advantage. I reported this area on my return to Washington.

After taking a good look at Whitefish Lake and drinking of

its clear waters, Freddie and I explored southward and found a sheltered nook where we built a fire to warm our snow-sodden feet as we ate our sandwiches. Returning over a high table of level glacier-polished rock to Fort Chimo, we stopped in at the Edmunds home, where Freddie's Eskimo mother—as gracious and agreeable a hostess as one could find in any land—served us tea with bread and butter. The more I saw of these natives of the North, the more I liked and admired them. In comparing Eskimos and white men, Desmond Holdridge well said, "For all our vaunted 'superiority' the most of us would die were we forced to cope with the Eskimo's conditions with his weapons. I cannot bring myself to regard a man who will attack and kill a polar bear with dogs and a spear and laugh and tell stories when tomorrow's food supply is still swimming, as an 'inferior.'"

By October 18 the *Cambridge* and the *Flow* were unloaded and their skippers anxious to be off. In addition they had provisions for only seven more days and it was at least a five days' journey back to Curling. Orders came through directing me to take the ships back to this Newfoundland port and with the *Flow* as my new flagship we bade farewell to Chimo.

At the river mouth the tide was now dead low. We tried to take the same course going out as coming in. What was my consternation when, 6 miles out from the river mouth, a snow flurry cleared and revealed a ledge standing well out of water and stretching across our bow scarcely half a mile away! We doubled back, then detoured cautiously around the east side of the ledge till we reached deep water. The reef must have been within half a mile of the course by which we had entered, but no sign of it had we seen.

On October 22 we entered the Strait of Belle Isle soon after midnight, coasted all day along the Newfoundland shores, and by nightfall were fast approaching the Bay of Islands where lay our destination.

Next day at Corner Brook I left the trawlers and made my way home to Boston, where I learned to my surprise that in spite of all the business of being sworn in as an Arctic Pilot by the Civil Service, I had been on active Naval duty all the time. Orders had come to Squantum from Washington directing me to report to the South Boston Army Base, but they could not be delivered, for at the moment I was navigating a fleet of trawlers across Ungava Bay.

As the first man to return from any of the Crystal groups I deemed it appropriate to report in person to Colonel Craig in Washington. Arriving there I learned that my appointment to the Civil Service was "wholly illegal" and jail sentences seemed to hang over the heads of all concerned. But a wise lady clerk was found who knew how to cope with such things. She simply had the appointment canceled, and we all breathed freely again.

Plotting all the soundings I had recorded, readjusting the position of the mouth of the Koksoak River in the light of our observations, and reporting these things to the Hydrographic Office in Washington occupied me through all of November.

On December 7 the attack on Pearl Harbor roused the nation from its long, deep sleep.

PART TWO

- 1942 -

6 Captain Bob Bartlett and His Schooner

By the middle of January, 1942, I had tried every possible way to bring my training in physiology to bear on wartime needs, but in vain. Then I found a billet to my liking. I was attached to the Hydrographic Office in Washington to help revise old charts of Labrador and to plan new ones for Naval aviators. It was unheard-of to assign a medical officer to such duty, but many such things were being done in this way. Through contacts with the Army Air Corps I dug out a wealth of recent aerial photographs unknown to the Navy, showing details of the Labrador coast, that enabled me to replace old charts based on guesswork with new ones portraying truly the intricate island archipelagos lying off the coast.

From Charlie Hubbard who was now attached to the Ferry Command of the Army Air Corps I learned much that was going on. In ski-equipped planes he had visited all three Crystal bases, where the garrisons were recording weather conditions. Farther south at Goose Bay, the Canadians had already made three landing strips, forerunners of a great airport. And momentous developments were going on across Davis Strait in Greenland.

During these months I often thought of the problem confronting the navigators of ships that would bring men and machinery up to Chimo to build the new airport at Crystal One. The uncharted reef, covered at high tide, that had risen ominously across our path as we came out in the trawler *Flow,* might well wreck ships and bring disaster to the enterprise. In April I submitted a memorandum to the Hydrographer suggesting that a survey ship be sent with or preceding the next group of supply ships going to Crystal One. This survey ship should be equipped to wire-drag and buoy the channel selected as most suitable for crossing the bar, and if this plan proved unfeasible, the ship could stand by to pilot successive supply ships and to explore and chart the bar and river mouth.

On June 3 I was called to the War Department for a conference with Colonel Bruton of the Engineer Corps, Major Ely, and Captain Grening on this very problem. The best existing chart of the Koksoak River was on a scale of about one to two million, the great river appearing as a tiny thread of ink, too small to show details even had they been known. The photographs taken from the plane in which Elliott Roosevelt and I had flown the previous July showed much of the river's shoreline, and my record of fathometer soundings, as the trawlers passed up and down the river, could be fitted into this outline to make a chart of sorts, not accurate but better than nothing. Grening asked me how soon I could produce such a chart for the contractor, to which I replied rashly, "Friday, June 5." Working feverishly far into the night, I sketched a map that gave some idea of the river and delivered it to Mr. Johnson in New York. As the head of the Al Johnson Construction Company, which was contracting to build the airports at Crystal One and Crystal Two, he was already concerned with plans about shipping equipment into the Koksoak River.

I was then told that I was to go north with Captain Bob Bartlett of North Pole renown on his famous schooner *Effie M. Morrissey,* to carry out the sort of survey of the approaches to Crystal One that I had urged in my memorandum. Admiral Bryan, the Hydrographer, approving the plan to detach me for this temporary duty with the Army Engineer Corps, asked me to get as much mapping material as possible for the Hydrographic Office.

"Captain Bob" and his beloved fifty-year-old schooner *Morrissey* are by-words to all who know the Arctic and her icy waters.

Bartlett, a native of Brigus, Newfoundland, was destined from birth to a lifetime of Arctic expeditions. His great-uncle, Isaac Bartlett, headed the relief expedition which rescued the Tyson Polar party from drifting ice floes sixty years before Peary's discovery of the North Pole. Three uncles commanded previous expeditions of Admiral Peary, and Captain Bartlett himself made his first northward venture in a vessel commanded by an uncle in 1897-98. In 1905, when Peary was planning the *Roosevelt,* the ship which Bartlett commanded on her epic trip to the North Pole, young Bartlett aided him in designing the vessel. Although only twenty-eight at the time, he was given command of the final party and reached a point within less than three degrees of the Pole. The youthful skipper shared in the honors heaped upon Peary. Later he stood stanchly by his leader in the controversy that developed over the claims of another explorer, Frederick Cook.

Perhaps his most thrilling adventure came in 1913-14 as an outgrowth of his trip as commander of the *Karluk,* aboard which Vilhjalmur Stefansson went to the Arctic through Bering Strait on a Canadian exploration trip. Foreseeing that the vessel would eventually be crushed, Bartlett unloaded enough provisions for two years and left them on the ice. One hundred and twenty-five miles from that point the vessel was splintered to bits. The skipper led seventeen men, an Eskimo woman and her two children to Wrangel Island where the cache of food had been left, and with an eighteen-year-old Eskimo boy, set out on foot for help. The pair traveled over treacherous shifting floes to the Siberian mainland, 100 miles away, and trekked down the coastline 1,000 miles before they reached Emma Harbor. Later Bartlett piloted a United States revenue cutter from St. Michael, Alaska, to the place where he had left his men.

Bartlett was given command of the third Crocker land expedition to North Greenland in 1917, when he rescued Donald MacMillan, who had been north for the past four years. World War I activities, surveying Alaska for air bases, summer junkets for museums or with students, and writing three books, all tended to confirm Bartlett's reputation as an Arctic "specialist." The intrepid explorer often expressed amazement at the popular idea that the Arctic was a hazardous place. He denied that northern expeditions were dangerous and full of strange hardships and said they couldn't be compared with tropic expeditions.

"Those who come through the tropics alive are usually wrecked by fever," Bartlett once declared, "While look at us—we come down from the Arctic in perfect health. We've had a fine time and it's good for us." Obviously the United States was more than fortunate in having such a man to aid in the subarctic projects of the Army and the Hydrographic Office from 1942 to 1945.

The *Morrissey,* which had served Captain Bob as home, office, and magic carpet for some twenty-two years and 300,000 miles, was a two-masted, 98-foot schooner, a product of the Essex, Massachusetts, yards. Her hull was sheathed in ice-withstanding greenheart and more than once Bartlett owed his life to her seaworthiness. Although always a confirmed bachelor, the skipper cherished his *Effie Morrissey* as a wife. Colorful, vibrant, fiercely independent, and at times explosive, Bartlett was a man who made a deep impression on me.

On June 9, 1942, I made my first visit to the *Morrissey* lying at the Army Base in Brooklyn. The sturdy vessel had her decks piled high with a vast litter of gear of every sort, just brought aboard in preparation for an Arctic voyage and awaiting stowage. Bartlett welcomed me cordially and listened with interest when I told him of our plans to go to Ungava Bay and survey the entrance of the Koksoak River.

In spite of all the gear heaped on the *Morrissey's* deck there were no surveying instruments or buoys to mark the channel when we had sounded it. These items were of utmost importance and were soon supplied by Captain Torning, an officer on the dock who proved to be a veritable Santa Claus. Additional supplies included wire-drag gear for sweeping channels and a range-finder for measuring our distance from the shore. The *Morrissey* was already equipped with an echo sounding fathometer.

As I was discussing plans with Captain Bartlett, an Army officer entered the cabin and said, "Bob, I'm giving you a couple of machine guns."

"Fine!" said Captain Bob, "We'll use 'em on walrus."

"Those aren't for walrus, they're for submarines," said the officer.

"To hell with submarines!" said Captain Bob. "We'll use 'em on walrus."

"How do we go north?" I asked.

"We're going by way of Brigus. I've got to pick up a whaleboat I left there," was the gruff reply.

Brigus is a picturesque town nestling among the tall, rocky cliffs at the head of a deep cove in Conception Bay, some 30 miles northwest of St. John's, Newfoundland. The journey from New York to Labrador is nearly 300 miles longer by way of the outer coast of Newfoundland than through the Gulf of St. Lawrence, but Brigus is the home of all the Bartletts. Here in a charming house lived Captain Bob's aged mother and his sister. All of the *Morrissey's* crew were Brigus men, and all had families there. You may be sure that at least a whaleboat was left there to be picked up. The *Morrissey* could no more go north without a rousing Godspeed from the assembled populace there than the moon could rise in the west. When I saw Brigus and all that it meant to those good people I was more than reconciled to the digression.

The *Morrissey* now ranked as an Army Transport and was under the jurisdiction of the North Atlantic Division of the Engineer Corps, commanded by Colonel Woodbury, with headquarters in New York City. They were eager to have her sail north as soon as possible. Captain Bob said he would be ready to start by June 17. But great as is the task of equipping a large schooner for a summer's surveying in the Arctic, it is as nothing compared to the task of getting orders authorized, drawn up, signed, and delivered.

My small part of that job comprised thirteen long-distance telephone calls and 1400 miles of travel by train and airplane.

The *Morrissey* would take at least a week to reach St. John's, and I was to go there by airplane; thus, when assured she would start without further delay, I had several days of respite. I therefore took a three-day vacation from all the turmoil by going to Rehoboth, Delaware, and flying with the Civil Air Patrol out over the waters off Cape May. Skimming low over the sea, we examined the wreckage of ships recently torpedoed by enemy submarines and searched for survivors or victims. The grim picture of havoc close to our shores was appalling. In many ways these small planes rendered gallant service; they spotted some U-boats in time to summon bombers that crippled or sank them, their presence kept others submerged and thus unable to strike, and they effected the rescue of survivors.

7 From Brigus to Port Burwell

It took a week or more to cut through the barriers of red tape, plane priorities, and official abracadabra that stood between me and St. John's, but at last these hectic and, at times, panic-filled days were behind me. The night of July 5 found me comfortably housed in the sumptuous new officers' quarters of the U. S. Engineer Corps at Fort Pepperell, an enormous Army Base under construction on the shores of Lake Quidi Vidi just outside St. John's. Here I found my friend Mr. Preston of the reconnaissance flights to Northwest River and Chimo, ensconced in the office of Major Hiatt, the officer who was to deal with our expedition from here on.

From Hiatt I learned that the Norwegian trawlers *Polarbjorn* and *Polaris* had just left for the Koksoak River and Crystal One, taking with them Mr. Balkin, the engineer for the Al Johnson Construction Company. Hiatt also said that after we had done the most urgent sounding at the Koksoak River with Balkin, we should probably be sent north to Frobisher Bay. At that I commenced a study of Major Crowell's report on the geography of that intricate and ill-charted sheet of water. Crowell had established the Crystal Two base on an island far up the bay in October and there he had spent the winter as commanding officer of his small garrison of ten men. His report, though sketchy, was of value to anyone who would try to find him.

On July 7 a dispatch came from New York telling us not to go to Crystal One at all, but instead to meet the trawler *Polaris* at Port Burwell and go thence with her direct to Crystal Two. Apparently they felt that they already knew enough about getting into the Koksoak River, and exploration of Frobisher Bay was much more urgent, especially since they were still in doubt which of three sites there to use for an airport: (1) Captain Roosevelt's choice, discarded by Crowell's party as having no good anchorage; (2) the island where Crowell had made camp and wintered; or (3) the area east of Sylvia Grinnell River at the head of the bay, favored by Hubbard and others, after they had viewed it from the air in June. Our job would be to see if site Number 3 could be reached by sea, and if so, whether Hubbard was right in thinking it would serve better than the island for an airport. We should then have the task of surveying, sounding, and charting the approaches to whatever base was chosen.

We had scarcely adjusted our minds to the change of plan when word came that the *Morrissey* had arrived at St. John's. We motored immediately to the city, arriving at the waterfront just in time to see the *Morrissey* gliding by down the harbor. We pursued her to the wharf we took for her destination, but when we got there she was already heading out of the harbor, presumably bound for Brigus.

We had much gear and six men to go aboard her, and Hiatt wanted to confer with the Captain. For these reasons I felt I should halt him if possible. A U. S. Destroyer was tied to the wharf. I hurried aboard and explained hastily to the O.O.D. what was up. He called a motorboat on the "double" and sent me off with two sailors to catch the *Morrissey* if I could. The boat was only good for 7 knots, the *Morrissey* for 5 knots under power. When we reached the harbor mouth the *Morrissey* was a good mile away with foresail and head sails set to a fair breeze and going well. Nevertheless we took up the chase and soon saw we were gaining. After a chase of about half an hour, the *Morrissey* apparently noticed us and luffed up into the wind; soon we were alongside.

I told Captain Bob why we wanted him at St. John's. He replied in his usual trenchant style that he had tried to reach the Army officers by phone, but those he knew were away and the telephone lines were busy. He didn't know I was there. He had a lot of stuff to put ashore at Brigus to make room for the men and gear we were bringing, and he had been told that everything could be sent to Brigus by truck. The suggestion that he should return to St. John's after laying his course for Brigus didn't go well with him at all. I yielded (the only course with him) and said I would explain matters to Major Hiatt. The *Morrissey* squared away again for Brigus, and we returned through the dramatic gateway in the high cliffs that guard the entrance to St. John's.

Next day Captain Bob motored over from Brigus and called on the engineering officers at Fort Pepperell for a conference, if it could be called that. Rather it was a stirring eulogy of his former chief, Admiral Peary, and a dissertation on provisioning ships for the north and on the family men who made up his crew. But he promised to be ready to sail on the morning of July 10. On that day all of us who were to go north with Bob motored over the winding road to Brigus. My fellow passengers were E. Drahos, representative of the Al Johnson Construction Company, and a group of surveyors led by Henry Kadjan. I was invited to an early dinner at the Bartlett home, where I met the Captain's mother, only a month short of ninety but clear-eyed and alert, a wonderful old lady, duly proud of her distinguished son.

At 3:30 that afternoon the *Morrissey* got under way. The people of Brigus, standing on the wharf, gave a traditional three cheers, and we stood out of the harbor into a calm gray sea where patches of fog lingered after the rain of the morning.

The crew of the *Morrissey,* besides Captain Bob, consisted of his brother Will Bartlett, Mate, and the following veterans of many an Arctic cruise: Leonard Gushue, Engineer; George Pomeroy, Second Engineer; James A. Youden, Boatswain; Ernest Whalen, Cook; Douglas Bishop, Assistant Cook; George Bartlett, Carpenter; Reginald Wilcox, Able Seaman; Bill Fowler, Able Seaman; Samuel W. Bartlett, Able Seaman. In addition, two men signed on for the summer's operations: Dr. E. C. Rigby, Ship's Doctor; and Corporal Majer, U.S.A. Radio Operator.

For eight days we traveled steadily north by the long east coast of Newfoundland and the longer coast of Labrador. A steady 5 knots was our speed under power and 7 knots or so when a good breeze filled the sails. Captain Bob pointed out two large, flat-topped icebergs as probably coming from Baffin Island. They were very large and low with level surfaces, thus contrasting with the tall Greenland bergs with their irregular, fantastic shapes, far more numerous in the Labrador current.

As we neared Cape Chidley, the salient peaks of the Torngat Range, accurately placed on Miller's map, afforded a perfect opportunity for precise cross bearings with the *Morrissey's* gyro compass, thus telling us just where we were. Late in the afternoon fog banks closed around us and hid Cape Chidley from view. Toward it we drove on under power over a calm gray sea with no landmarks in sight. Straining my eyes, I gazed into the fog, looking for the cape. A dozen times I thought I saw it, then realized it was but a trick of the imagination; nothing was in sight but a gray void. At last a steep cliff with a snowbank loomed up through the fog to port, then another cliff dead ahead. We came close to the bold shore to keep it in sight for guidance, till a sudden strong gust of wind from the north and a powerful eddy seemed to drive us dangerously toward the base of the cliff. Swerving out, we cleared the point. Port Burwell was one of the few places in the north to which Captain Bartlet and his crew were strangers, but as I had been there twice I was permitted to pilot us in. The water was turbulent with tidal currents. A giant whirlpool formed right beside us and in a few seconds a great block of ice, much bigger than a man, was sucked down and disappeared into the spiraling vortex. Gushue, the engineer, standing on deck, shook his head and exclaimed, "A good-sized rowboat would have gone down in that."

In the latitude of Cape Chidley in July twilight lasts all night, but now the clouds were dark and murky, and the dusk of night had settled on Gray Strait. Keeping well offshore to clear the reef that lies off Flat Island, we lost the shore in the haze and darkness. The Captain turned in and left the Mate and me to grope along the west shore of Killinek Island. At the darkest part of the night we picked up the shore close to Port Burwell; the Mate stopped the engine and let the ship drift until daylight.

The winter snow was mostly gone from the hills, but a chill wind still blew in from the icy waters of Ungava Bay as we dropped anchor close to where the *Fabia* had lain last year. There was a large iceberg on top of the *Sicilien's* anchorage. The surveyors came on deck and scanning the bleak shoreline, asked, "Where is Port Burwell?" "This is it," I replied. The surveyors could hardly believe me, for all they could see was rocks, grass, and lingering snowbanks, when they expected docks, warehouses, waterfront saloons, and other tokens of civilization.

Despite Major Hiatt's positive assurance that we should find the *Polaris* awaiting us, she was not there. Should we wait, or should we proceed to Frobisher Bay without her? The question lay on the table that day.

A small sloop lay at anchor in the cove close by. Aboard her we found Mr. Anderson of the Hudson's Bay Company with a crew of three Eskimos with whom he had just come from Chimo. He had seen the *Polaris* at Chimo and was surprised that she was not here. We discussed the problem and decided that our mission to Crystal Two was so urgent that we should go on our way with no further delay. I wrote a letter to the Captain of the *Polaris* telling him to follow us and that we would look for him in Frobisher Bay, east of Pugh Island, and took measures to see that it would be delivered. Captain Bob concurred and decided to start north next morning.

8 Frobisher Bay

Frobisher Bay is a deep inlet, thrice the area of Long Island Sound, that penetrates Baffin Land for 170 miles in a northwesterly direction from lat. 62° N. Forty-five miles wide at the mouth, it tapers to fifteen miles near the head, where it receives the waters of the Jordan and Sylvia Grinnell Rivers.

This Bay was discovered in August 1576 by the famous Elizabethan navigator Martin Frobisher, who was looking for a Northwest Passage to the Indies. He was certain he had found a northern counterpart for the Straits of Magellan. A brawl with the native Eskimos and the loss of his ship's boat forced Frobisher to return without penetrating the Bay far enough to discover that he was wrong. Some samples of iron pyrites that he brought home were assayed in England as gold ore, and the features of his one Eskimo prisoner were thought by the learned to be proof that Baffin Land was a part of Asia. So the "Company of Cathay" (China) was organized, Frobisher returned to his "Strait" in 1577, pushed a little

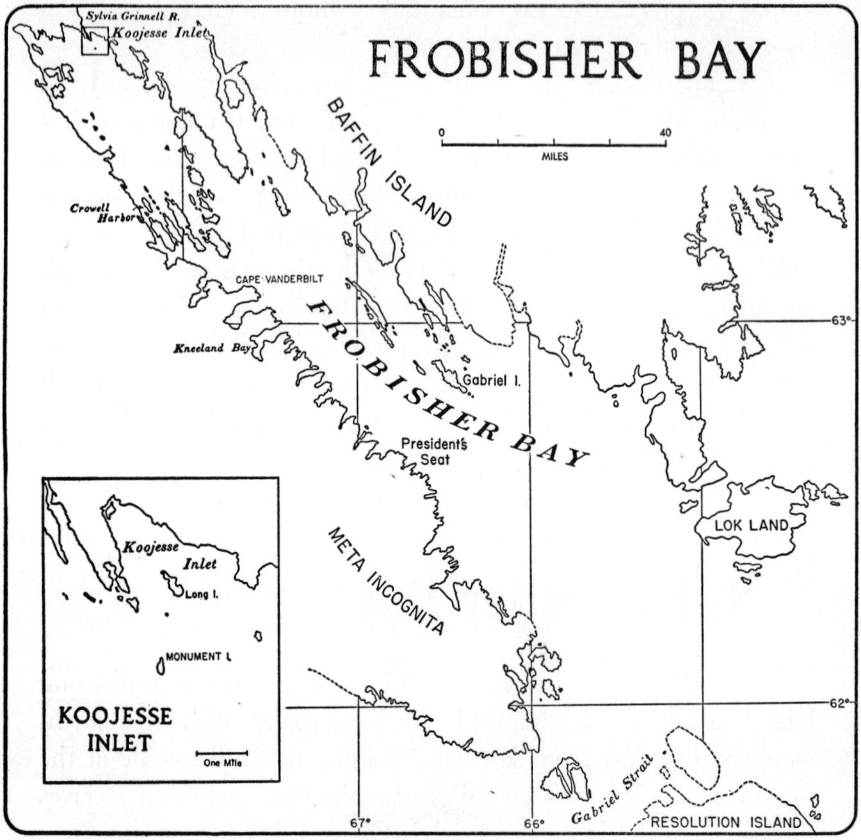

farther up the Bay, and brought home 200 tons of iron pyrites, for which he was given a present of £100 by the Queen. Again, the English metallurgists pronounced the iron pyrites to be gold-bearing, and Frobisher made a third voyage in 1578 to obtain more of the same, in command of a fleet of fifteen vessels. This time he made a temporary settlement on Kodlunarn Island, set his men to work mining, and took home several shiploads of the worthless iron ore. The trenches where he dug for gold, and the stone foundations of the huts that he built on the island were seen by Charles Francis Hall in 1862 and by Dr. Duncan Strong in 1927. An artificial canal, 150 feet long and 15 to 20 feet deep, that Frobisher's men dug either as a dock for repairing small craft or as a convenient place to load the ships is still visible.

Needless to say, the ore turned out to be worthless. It was "cast forth to mend the highways" in England; the Company of Cathay folded up after spending over £20,000, and Martin Frobisher found employment in the Royal Navy.*

The first white man to explore the bay up to its head was Charles F. Hall of Cincinnati, who went to Baffin Island in 1860 to search for traces of the lost Franklin expedition. He remained there till 1862, living with the Eskimos and with them traversing the length of Frobisher Bay by whaleboat in summer and by dog sled in winter. The Eskimos knew well enough that it was a bay, not a strait, and Hall verified their statement by his own observations. In 1865 Hall published a detailed account of these journeys in a large volume entitled *Arctic Researches and Life among the Esquimaux*. This book included a map of Frobisher Bay which, in view of the area covered and the crude means at his disposal, is remarkably good. Hall's map of 1865 had been copied without change in all Government maps and charts of the area right down to the time of our visit in 1942. Although it shows quite truly the position, size, and general shape of the bay, its portrayal of the most important groups of islands is so inaccurate as to be extremely confusing.

Hall devoted special attention to learning all he could about the visits of Frobisher nearly three centuries earlier. He questioned many Eskimos living in various parts of the bay, especially the older ones. A veritable fountain of ancient lore was found in the person of an aged woman named Ookijoxy Ninoo whom he judged to be at least a hundred years old. All of these authorities told him just the same story: Kodlunas (white men) came long, long ago in three successive years, first year two ships, second year three ships, third year many ships. This story, handed down from

* Notes by S. E. Morison from Vilhjalmur Stefansson, *The Three Voyages of Martin Frobisher* (London, 1938), especially vol. I, Introduction, and vol. II, pp. 240-247.

father to son and grandchild through many generations, agreed perfectly with written history which records two ships in 1576, three in 1577, and fifteen in 1578. According to Hall, Frobisher's last fleet was sent by Queen Elizabeth to found a colony. Written history records that five of Frobisher's men were lost. Numerous Eskimos told Hall that five white men who were detained by Innuits (Eskimos) dug a sloping trench close to the water's edge and in it built a boat with lumber brought in the ships for building a house; then sailed away, only to perish from the cold. The natives showed Hall the island (Kodlunarn, named for the white men) where all this happened. He paid several visits to the island, found tile, bricks, iron, and coal brought from England, also the ruins of three stone houses and the ship's trench by the shore. An engraving in his book shows this trench in detail, and it was with a real thrill that I recognized and photographed it on a subsequent trip.

There are three main entrances to Frobisher Bay, or Tin-nu-jok-ping-oo-se-ong, as the Innuits called it in Hall's day. There is the wide entrance northeast of Resolution Island, almost as broad as the bay itself; west of Resolution Island are two wide passages, one on each side of the Lower Savage Islands, a large cluster lying midway between Resolution Island and East Bluff at the tip of Meta Incognita, the southernmost peninsula of Baffin Island. The eastern of these two is called Gabriel Strait. With his characteristic independence, Captain Bob used none of these, but pioneered his way through a narrow passage, probably new to white men, between Resolution Island and an outlying island off its western shore.

The charts show Resolution Island as one large island with a suggestion of minor satellites. Our Navy planes flew over it in 1944 and brought back photographs showing the great land mass to be a labyrinth of islands riddled with inside passages. But this was entirely unknown to us as we approached from the south. Before noon on July 20th we sighted the south point of Resolution

Island. Near it loomed the Canadian Government radio towers that form an important link in communication in this lonely region and render great service to ships in thick weather by locating them with the direction finder. We coasted along the southwest shore toward Gabriel Strait, but on reaching the western corner of the island, found that wide passage choked with fog. It was in this situation that our skipper unhesitatingly steered into the narrower but fog-free passage inside the islands, relying on his masthead lookout to warn him of rocks ahead. The passage was deep all the way, and late in the afternoon we emerged into the great bay named after the dour old Elizabethan navigator.

The next morning we entered our first ice floes, a picture of rare and thrilling beauty; the ice pans either of clear blue or snow white, in strong contrast against the calm water reflecting the gray cloudy sky. Many small birds, identified as murres, disported themselves in the sheltered pools among the pans, although the temperature of the water was two degrees below freezing. Far to westward appeared the high land of "Meta Incognita" so named by Queen Elizabeth I, on the southwest side of the bay, surmounted by the broad snow fields of Terra Nivea and Grinnell Glacier, twin icecaps shining in the sunlight.

Meta Incognita is associated with a tale of supernaturalism and eerie coincidence. Eleven years previous, just as I was about to start on the 1931 Labrador survey, I was visited by a carpenter who lived in Boston. He told me that within the past three months he had begun to receive messages and visions from spirits. On a recent evening they had come to him with a vision of an airplane getting out of control. At first he thought it was a stunt flier; then it was further revealed to him that the plane was that of Nungesser and Coli, two French aviators who tried to fly the Atlantic westbound and were never seen again. The vision revealed a plane going into a dive over the ice as it approached the Atlantic Coast and coming down on a narrow strip of land called Meta Incognita. The carpenter told me he was given to believe that this land was just north of Hudson Strait and south of Baffin Island. At my request he made a diagram of the location as he saw it in the vision. Spending nearly a day at the Public Library studying maps, he finally found Meta Incognita, a narrow strip along the south shore of Baffin Island. He went on to say that there was a Great Crevice which now marks the course of Spirit River, which is in both Baffin Island and Labrador. I asked him how the same river could flow across the salt water of Hudson Strait and he replied that he understood it to be the course of a river in some previous geological age when it was all one land. He said that "they" (the spirits) told him the wreck of the plane would be found on Meta Incognita between the Great Crevice and Resolution Island. Spirit River, he believed, was a legend among the Indians and that if the right persons were found they would know all about it.

The visions further revealed that the plane's wings were white. Subsequent inquiries at a newspaper office proved that, indeed, the whole plane was white! In addition the man told me he had never seen a map of, nor studied the geography of this region prior to his visions, yet his description of these parts tallied with reports made by MacMillan after a trip in 1929 on which he explored the Great Crevice on foot. Most of the charts previously in existence had shown only one icecap, but photographs taken in August 1942 from the seaplane carried by U.S.S. *Bear* showed a wide gap between the Grinnell Glacier and Terra Nivea, with the Great Crevice traversing it. One of these is shown in this book.

One may conjecture that the sudden appearance of the unfamiliar name "Meta Incognita" in the carpenter's daydream was a message from spooks or from some other source via an occult process of telepathy, or one may conclude that he had at some earlier time seen the name on a map and forgotten it, and that it was neatly tucked away among his unconscious memories. His knowledge of the Great Crevice is indeed surprising; perhaps it came to him from MacMillan via the grapevine between 1929 and 1931. The fact that the wreckage of the plane was never found did not detract from the eeriness of the carpenter's tale, for Meta Incognita is a vast area and a desolate one and may to this day hold a grim secret buried in its icy depths.

Now we had other things besides Nungesser and Coli to think about. Entering the ice floes we dodged the larger pans and steered a more and more tortuous course as the pack ice became thicker. It was an old story to this seasoned crew. The man in the barrel at the foremast head picked the lanes through the ice and directed the man at the wheel, and as he yelled, "Port" or "Starboard," the schooner zigzagged dizzily. When feasible, he followed leads of open water; when these didn't serve, the schooner simply rammed the pans with a jarring crunch that would have terrified a yachtsman and pushed them aside, leaving a streak of red bottom paint on the ice as it floated away on the quarter. These men knew what the ship could take and they let her take it.

Our sails were all furled, since the wind was scant, and the Captain preferred maneuvering under power in strange waters. Slowly and steadily we forged ahead up Frobisher Bay. Gabriel Island was sighted on the starboard bow, while to port the mountains rose higher and grander. Never have I seen such outlines! The Torngats of Labrador are higher, to be sure, and their fiords are as impressive as those of Norway, but even the Torngats do not present such an array of jagged teeth, stabbing the upper air, as this coast of Meta Incognita, with massive shafts of solid rock rising straight out of the sea to heights of well over 2,000 feet. Beyond these peaks and the countless fiords between them lay the smooth, white snow field of the Grinnell Glacier, sublime beyond the power of words to describe. Tongues of this glacier come down to the heads of some of these fiords and drop icebergs into the sea.

We were now close to latitude 63° North. It was a month after the summer solstice, which meant that the sun at midnight would be no farther below the horizon than it would be forty-five minutes after sunset in the latitude of Boston. As midnight approached and the twilight deepened, we had already passed the highest mountain on the shore to the southwest of us, named by Hall "The President's Seat" in honor of Abraham Lincoln, and were half way from Resolution Island to the head of the bay. Captain Bartlett stopped the engine and let the *Morrissey* drift on the deep, calm waters till 3:00 A.M. when daylight was returning.

9 The Search for Crowell's Island

Our first task in Frobisher Bay was to find Major Crowell's camp, the advance base of Crystal Two. Hiatt had told us that Crowell was on the lookout for us, but the radio message we tried to relay to him received no acknowledgment.

From the reports of Grening and Crowell we knew that Crystal Two was on an island near the southwest shore of the bay, about 20 miles northwest of Cape Vanderbilt, and about 8 miles southeast of the headland Captain Roosevelt had recommended for an airport site and had photographed from the air. These were our only clues.

The morning of July 22 dawned as bright and clear as one could wish. So calm it was that even in the icy waters of Frobisher Bay we were warm in the sunshine. Soon after six o'clock we passed a prominent cape that we rightly judged to be Cape Vanderbilt, for just beyond it began the complex island barrier that intervenes between the lower and upper bays. Captain Bob steered for a cove by the nearest point of Fletcher Island, where the barrier group begins. There, close to shore, we dropped anchor just before 7:00 A.M. and for a brief time feasted our eyes on the panorama of mountains and islands over the placid water, where fragments of winter ice gleamed white in the morning sunshine.

I offered to go through the islands in the whaleboat in search of Crowell's camp. Supposing it to be about 18 miles away, the 6-knot speed of the port whaleboat should get us there and back with many hours to spare before sunset. The Captain approved, and assigned Pomeroy, Fowler, and Sam Bartlett to be crew. Kadjan and Drahos decided to go too, for their first duty was to examine the field by the camp and see if it would do for an airport. Taking with us a light lunch and a pistol for signaling or other emergencies, the six of us piled into the whale boat and blithely set forth at 8:30 on our expedition.

In mystery stories we generally find one of two techniques. Sherlock Holmes, with his rare gift of penetration, surmised the answer long before he offered an intelligible clue to the reader, whose perplexity remained till the denouement burst on him in all its glory. Bulldog Drummond battled blindly with a hidden foe, sustained only by his spirit of adventure, while the reader, knowing the secret from the start, wished he could give the valiant hero a helpful hint. The second technique will serve better for the story of our hunt, and to that end the reader is referred to the two maps 6 and 7 printed side by side. The first is a copy of the part of Hall's map covering the area between Capes Vanderbilt and Rammelsberg. It was our basic chart, and though nothing on it corresponded at all to the description of Crowell's island, we surmised (and rightly, as we later learned) that Cape Rammelsberg was the headland that Captain Roosevelt had picked for the base and photographed. The other shows the same area correctly charted by the survey of 1943; and on it are shown the *Morrissey's* anchorage, whence we started, and the course that led in time to finding the long-sought camp. Thus the reader has before him a key to the situation—one that we should have given much to possess.

Steering northwest, with Fletcher Island to starboard, we had gone about 2 miles, when looking back, we saw a ship approaching that turned out to be the *Polaris*. With nearly twice the Morrissey's speed she had been overtaking us all the way from Port Burwell. Mr. Balkin was aboard with some lumber and carpenters to start

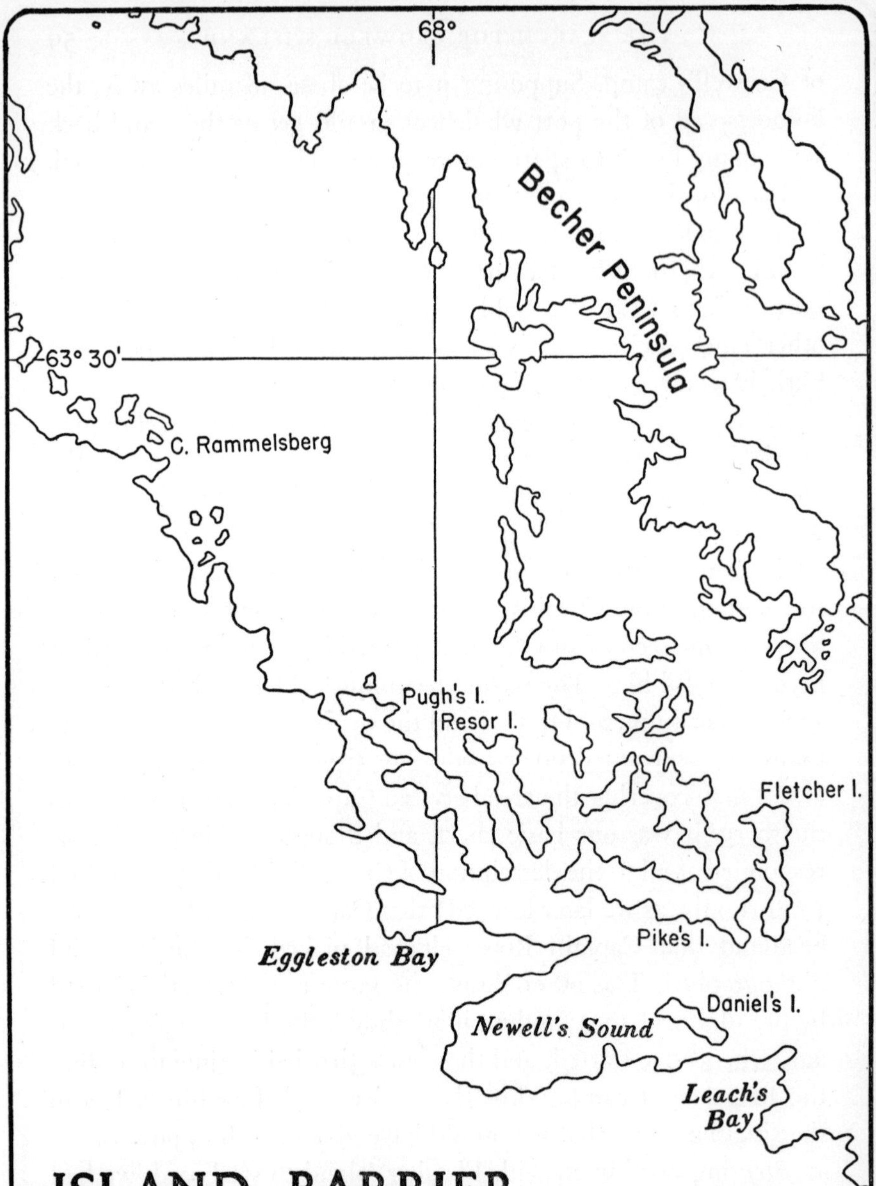

ISLAND BARRIER
HALL 1865

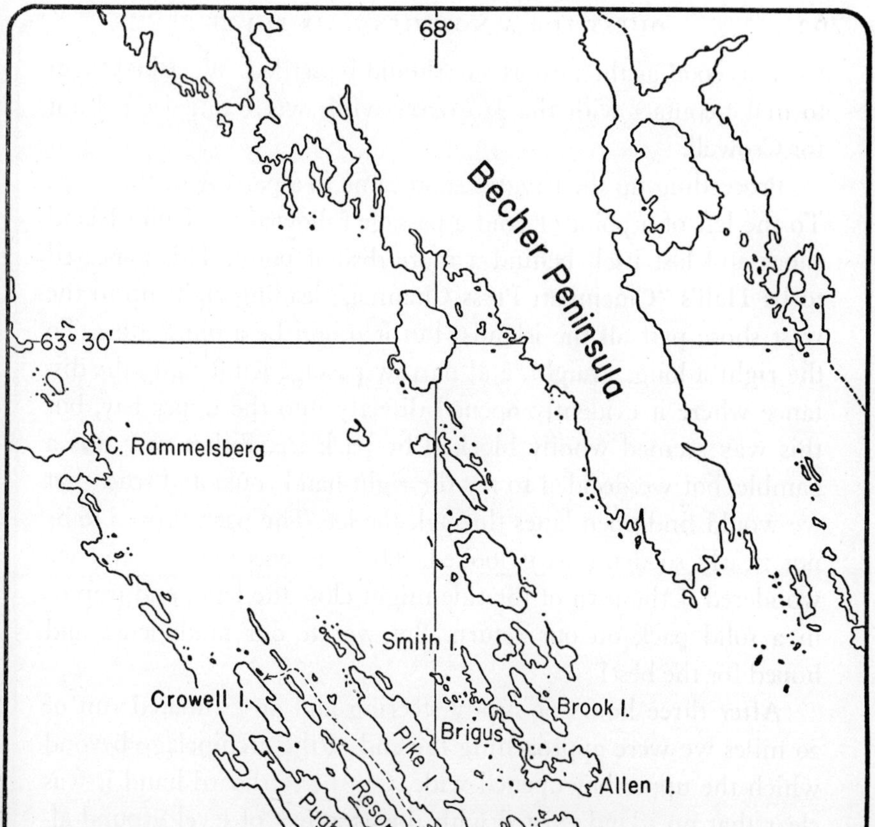
68°
63° 30'
Becher Peninsula
C. Rammelsberg
Smith I.
Crowell I.
Brook I.
Brigus
Pike I.
Resor I.
Pugh I.
Allen I.
Fletcher I.
Eggleston Bay
Whaleboat passage
Morrissey at anchor July 22,'42
10 Miles
Newell Sound
Daniel I.
Leach's Bay
ISLAND BARRIER
H.O. SURVEY 1943

work as soon as the airport site should be settled. We asked them to make contact with the *Morrissey* while we resumed our hunt for Crowell.

Proceeding up the bay, we soon came to a parting of the ways. To the left of a point of land a passage followed the Baffin Island shore and lost itself behind a more distant point. This appeared to be Hall's "Cincinnati Press Channel," leading right up to the west shore past all the islands; but it might be a blind alley. To the right a long, straight, and narrow passage led far into the distance where it evidently opened directly into the upper bay, but this way seemed wholly blocked by pack ice. Either way was a gamble, but we decided to try the right-hand route and trust that we would find open lanes through the ice. The pack proved to be not nearly so dense as it looked. The tide was with us and we wondered if the turn of the tide might close the lanes and trap us in a solid pack on our return. But we ate our sandwiches and hoped for the best!

After three hours of steady steering and an estimated run of 20 miles we were approaching the end of the archipelago beyond which the upper bay opened wide. On the starboard hand it was clear that no island existed with the 4500 feet of level ground alleged to adjoin Crowell's camp. To port the islands were larger, but high, rugged, and forbidding. One of my shipmates turned to me and said, "Where is this base we're looking for?" I replied, "I don't know where the base is, nor where we are, and the maps aren't any good. But I think the place must be somewhere in here." And I swept my arm over a wide and noncommittal arc to port. An opening appeared in that direction and through it we steered for the only island that looked large enough to offer a landing strip. Dense pack ice blocked our way and inspection from a height of land seemed our best bet, so we landed on the nearest island and started to climb the high hill that would command a view. Half way up the hill I was astonished to see in a nook among the rocks neat rows of slender shafts of different colored stone about 3 inches long, standing upright against the face of the cliff. "Children have been playing here," I exclaimed. It was an Eskimo dollhouse. Coming over the crest of the ridge we saw on the other side a large number of Eskimo dogs and other signs of a recently abandoned camp, but the Eskimos had gone away in their boats; only their dogs were left on the island.

To the northwest I recognized Cape Rammelsberg from my study of the photographs of it brought back by Captain Roosevelt. Nothing in sight offered a better hope of being Crowell's island than that which lay across the small, ice-choked bay to the west. But from this height it looked steep, rocky, and impossible for an airport. Examining its skyline with binoculars, my eyes at last lit on a small flag or rag on a staff protruding from the top of the highest ridge.

"There's a sign of life," I said; and at once we decided to force a way through the ice and investigate that island. After much pushing with the boat hook we gained the shore at about the time of the ebb tide. Sam Bartlett and Pomeroy stayed with the boat while the rest of us started for the ridge.

"Make it quick, or we shan't get out of here," they called after us, for the outgoing tide might well bring enough ice down the bay to jam the passage in a hopeless block.

We were soon rewarded by looking down from the summit on a broad stretch of level tundra, on the far side of which stood two houses and the radio tower of Crowell's camp. As we approached, Crowell and a soldier came out to meet us with high-powered rifles loaded and ready to repel an invading force. Happily, Crowell recognized me. He had received no word of our coming, nor had his radio operator heard our message of the day before. We took them wholly by surprise.

With the ebb tide gathering force every minute there was not a moment to lose. It was quickly agreed that Kadjan should stay at the camp and begin at once to examine the field, and that Crowell would return with us to the *Morrissey* to pilot her into the proper harbor when the ice would permit. We hastened back to the boat and shoved off. Pushing out again through the ice we regained the long passage by which we had come. The ebb tide had, as we feared, caused a great congestion of ice pans in the narrowest part. Four lanes we tried before finding one by which we could penetrate the pack. Crowell looked about the small, open whaleboat. "Are you fellows fond of eating shoelaces?" he asked. "No, why?" "It looks as if you might have to," he replied.

We had rashly eaten everything that we had brought from the *Morrissey,* and we were getting hungry again. But we found an opening at last and reached the schooner after a ten hours' absence. After a welcome supper the *Morrissey* weighed anchor and with Crowell's expert pilotage, threaded her way into Cincinnati Press Channel through scattered floes till the twilight grew dim. We came to anchor near the *Polaris,* just at midnight, in a sheltered harbor, Hall's Eggleston Bay.

The next day, which was clear, calm, and warm, we entered the narrowest part of Cincinnati Press Channel. Here, through the clear water, we could see the bottom 4 fathoms deep. Proceeding cautiously, we soon gained the deep water beyond, and in another 3 or 4 miles came to the edge of a solid barrier of ice.

On the map it may be seen that north of Pugh Island there is a good-sized harbor between Crowell's Island and the mainland of Baffin Island to the west. This was the harbor where the trawlers anchored when unloading all that went into the making of Crystal Two in October. When Captain Bartlett saw the solid ice in the harbor in the west shore of Crowell's Island, he sought anchorage in a cove on the shore of Pugh Island where a projecting point offered protection from the onslaughts of the ice floe, in case it should break loose and charge down the passage on a strong ebb tide. That afternoon Dr. Rigby and I went ashore on the northern headland of Pugh Island and climbed to its highest point whence we could look over the many islands, large and small, that make the northwest fringe of the barrier labyrinth, and north of that the wide expanse of the upper bay, now a solid sheet of snow ice. The harbor by the base lay spread before us, and the ice that covered its surface looked like a lattice of snow ice with regular patches of open water. While on the hilltop we built cairns for markers, took photographs, and did a bit of amateur surveying from this excellent observation post before we returned to the ship.

I reported what we had seen from the hill and the appearance of ice about to break up in the harbor, but Captain Bob could judge ice better from the deck of his boat than we could looking down from a hilltop. There was no open water and he knew it, and in trenchant language he sounded a warning against venturing forth. "I've known a vessel to go into ice like that and get crushed!" he said in a marked manner. I offered no more comments on ice that afternoon. What had looked to us like patches of open water was really firm gray ice. We could only wait till the forces of Nature should break up the ice and sweep it out to sea.

10 Initial Reconnaissance; Koojesse Inlet

From the tale of the trawlers we knew that there were two passages at least by which ships could reach the upper bay. It was important to choose the best, and then chart it. In our exploration of "Whaleboat Passage," as we called the channel we had traversed between Pike and Resor Islands on July 22, we had noted shoal water extending some way off shore. The passage was 11 miles long and would need careful sounding throughout. Cincinnati Press Channel led more directly to the harbor, and less than 5 miles of it presented any likelihood of shoals. Crowell was convinced that it was the better route, yet we knew that at its narrowest point there was a shoal that called for study. From our sheltered cove at the north end of the passage we were well situated to commence this survey while waiting for the ice to make way for us.

We unpacked the fathometer,* rigged a frame to support it on the rail of the whaleboat, and began a series of runs down across the bar. We soon learned it extended right across the channel with depths of 10 and 12 feet at low tide. This meant that any freighter entering that way must await the tide, for few would draw less than 18 feet of water. But with a tidal range of more than 20 feet, and well over 30 on the spring tides, we could be sure of a safe margin at high tide. And Eggleston Bay offered a well-placed anchorage for ships awaiting the flood.

As we lay at anchor by Pugh Island making our daily sounding trips down to the bar, we settled into the routine of life aboard the *Morrissey,* a routine that prevailed with only minor breaks throughout the summer. Mealtimes on this famous schooner were memorable occasions. The table in the main cabin was not long enough to seat all the ship's company, and two sittings were required. All candidates for the first sitting were expected to be on hand promptly, but one of the most inviolable of unwritten laws was that on no account could anyone take his seat before the Captain emerged from the door of the engine room and sat down at the head of the table.

Conversation with Captain Bob was a privilege and a pleasure. In his notable career as explorer, matchless expert in the mastery of polar ice floes, staunch supporter of Admiral Peary and lecturer on his memorable voyages, he had come to know intimately many persons of distinction and eminence. From his rich store of experience, thrilling anecdotes were told, punctuated with wise and pithy comments on characters he had known and on life in general. Everyone joined in and there was always a lively and stimulating interchange of ideas.

---

*The fathometer is an echo sounding device that, when suspended in the water, automatically records on a long roll of paper the depth of the water.

I noticed that as we lay at anchor Reginald Wilcox, a member of the crew whose functions included photography and other things beyond the ordinary province of an "able seaman," frequently brought on deck some collecting gear and dredged for samples of marine life. From Captain Bob I learned that this was being done for the Smithsonian Institution. Thus our skipper was quietly doing something for the science of natural history on the side. Recently I have learned from the head of the Institution that Bartlett's contribution in this line was of real value.

Few skippers have ever been more considerate of their crews than Bob Bartlett. There was no military smartness about the ship, but morale was high, and the competence of the crew and their readiness for any task proved that he had the kind of discipline that makes men happy and at the same time eager and efficient. Captain Bob liked early hours; his favorite time for getting under way was 4:30 A.M. But by six o'clock in the evening the crew's day was done; then they were entitled to their rest, and only in the case of great urgency would he ask them even to put over the whaleboat or do a stroke of work. By seven o'clock he was usually fast asleep.

To a man of academic rather than seagoing habits, the resounding ker-chung, ker-chung of the power-driven windlass intruding in pleasant dreams at 4:15 A.M. was wont to evoke the muttered comment, "Why can't the old man keep reasonable hours?" But of course with equal justice he could retort with the same comment to the city man who burns the midnight kilowatts and sleeps till eight.

The most perfect picture of contentment in a veteran sailor was the gray-haired ship's carpenter with his old-fashioned drooping mustache. He seemed to belong to the ship as much as her mainmast or her rudder, his world wholly contained within the timbers of her hull and the sturdy bulwarks on her deck. Now and then he would glance up at a gleaming iceberg or a towering mountain peak, but usually he was completely and happily absorbed in his task, ever working with skilled hands and singing to himself as he worked. I believe he did not know or care where he was or whither we were going, so long as he was on the *Morrissey*.

Few indeed are the fields of human endeavor in which real manhood is displayed to better advantage than in good seamanship, and to have sailed with Captain Bob and seen the *Morrissey* crew in action among the northern ice floes is to have experienced the best.

Daily we ran the whaleboat from our anchorage by Pugh Island down Cincinnati Press Channel and scoured the bar with the fathometer in search of any pinnacle or rock rising above the prevailing depth of 2 fathoms. So swiftly did the current run over the bar that the whaleboat with her 6-knot speed could barely cope with it. This was the strongest tide we found anywhere in Frobisher Bay.

The land was not without its hazards, either. On July 26 during an afternoon's soundings we made a landing on the west shore of the passage where we needed another paint-spot to locate our soundings. The rocky slope was very steep, and 30 feet above the present low-tide level of the water a massive rim of ice (called ballycarta by the crew) clung to the surface of the rocks just below high-water line. Through a narrow gap in the ballycarta we climbed to the dry rocks above, and while Sam Bartlett painted the marker, I took sextant angles from it to our other marks across the channel. Returning through the gap in the ballycarta, I stepped down on the only available foothold within reach. It was a rounded dome of slimy rock. How Sam and Rigby had managed to balance on it I don't know. My foot slipped and I plunged head first down the cliff, fetching up against Rigby's legs. Fortunately he was on a solid shelf and stood firm. The box containing my sextant flew from my hand and bounced many times before it lodged just above the water. Watching it from my inverted position, I was so spellbound that I was unaware of a sprain and small fracture of the little finger, later revealed by X-ray. I was lucky indeed to have escaped without serious injury. The sextant, too, for all its bouncing was none the worse for wear.

That afternoon a stormy wind blew up from the southwest and heavy rain fell in the night. The storm served us well, for the combination of wind, rain, and a high spring tide in the night broke up the ice in Crowell's harbor and sent it packing out to sea. The next day dawned bright and clear and from our observation point on the hill top of Pugh Island we could see the harbor was a wide expanse of clear blue water, ruffled by the breeze, with only a few scattered ice pans gleaming white in the morning sunshine against the deep blue of the water. The *Morrissey* was soon under way and by noon we were snugly anchored in the long-sought harbor close to Crowell's camp.

Scarcely had we dropped anchor before a rowboat put out from the shore with Crowell and Balkin aboard. Kadjan's examination of the proposed field by Crowell's camp proved it to be inadequate for an airport of the size required. We must therefore proceed to the head of the bay as soon as possible and examine the site favored by Hubbard, lying east of the Sylvia Grinnell River and abutting on Koojesse Inlet.

A glance at maps 5 and 7 will show that the upper bay is cut off from the lower except for a number of narrow passages between the islands. It is a deep basin; hence, though the tide runs fast through all the narrow waterways, there is little current in the wide part beyond. For this reason the ice is slow to break up. But on July 30 it was evident that the ice sheet had broken and was fast moving out on the ebbing tide, aided by the several large rivers that enter the upper bay.

Accompanied by the *Polaris* we started our journey to the upper bay. Emerging from the maze of small islands into its wide expanse, we picked our way northwestward through the lingering ice floes and when open water permitted, steered for the mouth of the Sylvia Grinnell River.

About 6:30 P.M. the *Polaris* and the *Morrissey* anchored in 20 fathoms near a high island beyond which we could see the wide expanse of level land which was evidently the proposed airport site. Crowell and Balkin were eager to get ashore and examine the terrain and lost no time in mobilizing an Army tent and camping gear. The idea of spending the first night ashore at this important site appealed to me strongly and I tossed my sleeping bag into the lowered whaleboat. With us was Paluchi, an Eskimo patriarch of Frobisher Bay whom Crowell had taken on as guide, philospher, and friend, factotum and counselor soon after his arrival the previous October. The whaleboat took us in past the high island near the summit of which we saw a strange shaft of rock protruding vertically from the soil. We named it Monument Island. North of it was a small, sheltered bay that evidently was what Hall had named Koojesse Inlet (after his loyal Eskimo guide). The tide was high, enabling us to penetrate far to the head of the inlet where we landed on the rocky shore of a sheltered creek.

Before the sun set we had pitched our tent on a level bench of turf, moss, and wild flowers, commanding a beautiful view of the bay and the wide plain, destined soon to become a great airport. The mosquitoes, unbearably dense during the day, retired soon after sunset when the air got cool, and so did we, sleeping soundly in the tent till the bothersome little pests returned and roused us at sunrise.

Crowell and Balkin in their eagerness to explore the ground for airport development, breakfasted hastily on a slice of bread and a bit of ham and made off when the sun was barely above the horizon. I accepted Paluchi's offer of a cup of tea and then explored on my own, finding level ground extending far to the south and at one point penetrating the rocky rampart that bordered the west side of Koojesse Inlet.

As I explored the ridge I saw the *Polaris* and the *Morrissey* enter Koojesse Inlet and come to anchor in the excellent harbor

Captain Bob Bartlett surrounded by surveyors, doctor, and Commander Forbes on the MORRISSEY, *July 10, 1942, Mate Will Bartlett at the right (photograph by U.S. Army)*

MORRISSEY *leaving Brigus, July 10, 1942 (photograph by U. S. Army)*

MORRISSEY *bucking Frobisher Bay ice floes, July 21, 1942*

Ice floes in Cincinnati Press Channel, July 1942

Cincinnati Press Channel, July 27, 1942, MORRISSEY *and* POLARIS *at anchor*

Original Crystal Two camp, 1942 (established by Crowell in October 1941)

Trawler POLARIS *in Crowell Harbor, 1942*

Southwest shore of Frobisher Bay showing Delano Bay, 1942

View from anchorage in Kneeland Bay, August 1942

S. S. FAIRFAX *entering Frobisher Bay, August 23, 1942*

Meta Incognita, showing Grinnell Glacier, August 24, 1942 (photograph by U.S. Navy)

Great Crevice traversing Meta Incognita, August 1942 (photograph by U.S. Navy)

Eskimo camp, Sylvia Grinnell River, September 1942

Eskimos, Sylvia Grinnell River, September 1942

Development of Koojesse air base, September 16, 1942

that it offered. From all directions except the southeast it affords perfect shelter. Strong winds from that quarter are infrequent, but when they blow, the anchorage is a choppy place for small boats. I crossed the plain, climbed a high ridge to the west of it, and on the other side found the Sylvia Grinnell River. Half a mile from its mouth are white foaming rapids where the river makes its final plunge to sea level. On the bank was a beautiful patch of green turf where the Eskimos make their camp from time to time.

Returning to the ships I joined in a conference with Balkin and Kadjan. It was now clear that this was the place for the airport. The terrain was excellent, with level ground extending nearly the entire 6,000 feet needed for an ample runway. Bulldozers could easily level it as far as was needed. The harbor, readily accessible from the bay, offered a good anchorage for ships of any size that would come in. Kadjan and his assistants would at once begin their survey. The carpenters on the *Polaris* were prepared to build a camp in which the surveyors could live ashore, and my job was to begin a hydrographic survey of the harbor and approaches.

The standard method of placing soundings is to run the boat back and forth on closely spaced parallel lines, locating her at frequent intervals by two simultaneous sextant angles between three markers on the shore, the depth of water being continuously recorded by the fathometer. This requires organized teamwork for which we were not equipped. Instead we used a cruder method that proved adequate. The whaleboat would start from a well-placed point, close to the shore, and run at a constant speed, steering straight for another landmark. The fathometer was started at the same moment as the boat and traced its continuous profile of the bottom on the moving paper roll till the boat stopped at the end of the run. In this way the depth at any point on the course could be correctly placed on the chart. For added control, sextant angles on landmarks were taken as often as possible.

On August 1st our first task with the whaleboat was to land the surveyors on the beach to get on with their survey of the new airport. This done, we commenced our hydrographic work, zig-zagging back and forth across the harbor, and making runs out into the bay to cover the approaches as well as the anchorage itself. Sam Bartlett usually served as secretary, taking notes from my dictation as I wielded the sextant. Things happened so fast it was hard to keep the pace of dictation down to the speed of the scribe. I would hurriedly call, "Heading on course for north end of island; 10:52 A.M. mark on record, angle paintmark One to left tangent of headland 83°," at which point Sam would plaintively break in with "What was that?" and I would have to begin again at half speed. In time we learned abbreviations that could be recorded while we ran our courses. For many days this work went on, and the chart of the harbor grew.

The next weeks passed busily and pleasantly. The camp was built, the surveyors settled there and made a detailed survey of the field, and we on the *Morrissey* charted the harbor and explored the island labyrinth down the bay.

When occasional bad weather or engine trouble prevented our working outside there was plenty to do constructing the chart from our observations. The only place on the *Morrissey* where this work could be done was on the main cabin table, and even here it was by no means easy, for the steward insisted on starting to set the table at least forty-five minutes before each meal; two sittings followed, and then the steward had to eat his own meal in the galley before he would start clearing off the table. The Captain graciously offered me the use of the chart table in his own cabin, but that was chronically so littered with charts, almanacs, logbook, captain's spectacles, garments, pools of tea, and bread crumbs that I couldn't cope with it. In the midst of this chaos Captain Bob's logbook lay with its pages often open for inspection. The vivid account of our voyage in his picturesque language was worth reading, and sometimes helpful to me in checking the accuracy of my own hasty notes. One day I found a narration of my own doings in which I was referred to as "the O.G." What on earth did these cryptic letters mean? Perhaps I shouldn't have done so, but curiosity won out and I turned the pages back till I found the clue; he had me down as "the Old Gent," and this although he was seven years my senior!

11 Exploring the Bay; Arrival of Big Ships

On August 16 we had gone down the bay to check the latitude and longitude of Cape Vanderbilt, because a radio dispatch had indicated that cargo ships were coming soon to meet us there for pilotage up the bay to Koojesse Inlet. Captain Bob suggested that we scout northward and see if there might be a deeper, better passage into the upper bay than Cincinnati Press Channel with its two-fathom bar requiring deep-draft ships to wait for high tide before crossing. This proposal was greeted eagerly and we headed into a new part of the bay that none of us had ever seen, proceeding between the northeast shore of the bay and a long chain of islands marking the eastern border of the archipelago that separates the upper from the lower bay. We searched to the northward and sighted a high rounded knoll of reddish rock, looking for all the world like the photographs of the British Prime Minister's thumb held upright to signify the will to win. We named it "Churchill's Thumb" and so it appears on the Government charts today.

As we neared this landmark, we suddenly saw an opening between the islands leading right through into the upper bay. It was a narrow strait, but deep from shore to shore. From then on this was the passage of choice for ships bound for the new airport site; we named it Bartlett Narrows. Thence crossing the upper bay we passed near a large lone island with a façade of pinkish rock that inspired us to name it "Pink Lady." It was after nine when we anchored for the night in Crowell's harbor, well pleased with the discovery of what appeared to be the best way yet from the lower to the upper bay. Since morning we had made a complete circuit of almost the entire island barrier and had added much to our knowledge of its geography.

The U.S.S. *Bear,* famed for many exploits through a number of years, but chiefly for her part in Admiral Byrd's Antarctic explorations, was now serving as flagship for the Canadian Arctic Task Unit of the Greenland Patrol, with Lieutenant Commander Graves aboard in command of the unit. Frobisher Bay was part of his domain. Radio dispatches from her had advised us that on August 21 we were to meet a convoy with 350 men to build the airport. The orders directed us to Kneeland Bay, a fiord about 12 miles southeast of Cape Vanderbilt. To this fiord we went, and there behind a small island we found an anchorage to the Captain's liking, where a waterfall tumbled through a narrow cleft right into the bay, and above it towered a 2,000-foot giant fortress of solid rock.

On August 22, while awaiting word of the convoy, a party of us set out in the whaleboat to explore our surroundings. New ice had formed on the water, not owing to subfreezing air temperature, they told me, but because fresh water from the streams froze as it flowed over the salt water whose temperature was below 32° F. In places the ice was thick enough to batter the sides of the boat till the engineer slowed down to save the paint. Rounding the final curve of the fiord we came in sight of a lovely waterfall some 200 feet high, coming down through a vertical chasm at the head of the valley.

Near the entering stream the Captain saw some white whales, and suggested that we stalk them with rifles overland. All but he landed and walked over a fascinating terrain of mossy benches among the rocks. For some time we watched the white whales through binoculars, but they stayed on the other side of the fiord.

The battle of Stalingrad was now reaching its climax. All the world was watching breathlessly to see if that crucial stronghold would stand or fall. As we lay on a mossy bank in the peaceful sunshine watching white whales at play, the contrast struck me with great force. I said to Crowell, "This is a funny way to go about stopping Hitler from getting oil out of the Caucasus." He laughed, but added seriously, "It's the best way, though." So we thought and hoped. The outcome of the whole war hung in the balance then. If only the airport could be finished soon it might speed the flow of fighter planes to England and help them turn the scales.

Returning to the boat we stalked the white whales by water and then the fun began. We chased them to the west shore of the bay where the whole school of twenty or more presented a remarkable show, swimming fast just below the surface and, when winded, all coming up for breath at once in an almost solid mass of white backs. The color effects of their bodies emerging from the clear blue-green water were truly remarkable. They would get ahead of us when swimming under water, and when they came up for breath we would overtake and pass them, and then they would double back along the shore. Several shots were fired, without success, but both Rigby and Wilcox took yards of color movie film of them as they broached and dived again. This sport held us captive a long time, expending gasoline, gunpowder, and movie film, but we returned to the ship for lunch.

That afternoon the long-awaited radio came from the transport *Fairfax,* already entering Savage Strait and requesting us to meet and escort her up the bay. This brought us back with a jerk from the sport of chasing whales to the major task before us. At four o'clock we started eastward to meet the convoy, traveling at our usual 5 knots.

At 4 A.M. on August 23 I was awakened with the news that we had met the steamers. I hurried on deck and saw the sun, just arisen over a clear, calm horizon in the east, throwing a wonderful pink light on the snow-topped peaks southwest of us. We had turned around and were being followed by two big steamers, the *Fairfax* and the *Eleanor*. Our first convoy had been met, and a strange sight they were in this lonely wilderness. Crowell went aboard the *Fairfax* to pilot the convoy up the bay to the base. A ponderous package containing a communication from the Commanding Officer of the Greenland Patrol informed me that coöperation was being established between the North Atlantic Division of Engineers (Army) and the Greenland Patrol (Navy), and that I was designated by the Navy as communication officer and hydrographer on the *Morrissey*.

Busy days began for the *Morrissey,* transporting Crowell, as official pilot, to meet incoming convoys; not much time was left for hydrography. No sooner had Crowell taken in the *Fairfax* and the *Eleanor* than we received a radio dispatch calling for more pilotage. We retrieved Crowell from the base at 1:00 A.M. on the 24th and hurried down the bay to meet a tug with barges. We had put Crowell aboard her in mid-afternoon when we received a radio from the *Bear* saying that she was coming up the bay and that Commander Graves desired a conference with me. Soon the *Bear* appeared and I was sent aboard to pilot her the 25 miles up Cincinnati Press Channel to Crowell's harbor, and on the next morning to Koojesse Inlet. It was a joy to go aboard this famous ship with her trim appearance and atmosphere of efficient Naval organization. We soon left the *Morrissey* far behind struggling against the tide and anchored in Crowell's harbor in time for supper, and a grand supper it was. Then the ship's company gathered in a narrow, but adequate space and had movies, a new Army sound film. It was a good show and a great contrast to the life of the last six weeks.

Soon after breakfast the next morning the *Bear* started out the north entrance and made a quick run to Koojesse Inlet. There we found a busy scene; outside the harbor were anchored the *Fairfax* and the *Eleanor;* in her usual berth inside was the *Polaris,* now become a fixture, while the tug *Mathews* and another smaller one were playing around among the big barges, unloading supplies from the steamers. Here at this harbor, where less than a month before the icebound shores knew no sign of life but the seal, the sea trout, the eider duck, and an occasional fox, where probably no white man had ever trod, a turmoil of sea-borne traffic was already in evidence. The curtain was rising on the drama of landing vast stores for the building of Quonset huts, storehouses, and electric power lines, while high-ranking officers of the Army and Navy were examining the airport site and the preparations for its development.

The *Bear* carried a seaplane on her superstructure. The day before, this plane had made a long flight and taken two series of oblique photographs covering the southwest shore of the bay from East Bluff to Ney Harbor and from Cape Vanderbilt to Cape Rammelsberg. The skipper, Lieutenant Hinckley, now decided to launch the seaplane again and go up himself to make his own observations of the bay, taking a photographer to record some more of its geography. He invited me to go with him and select such views as would serve the purposes of our survey. Here was a golden opportunity to see the bay as one never can from the surface of the sea, and to search the clear waters for dangerous shoals. An unusually low spring tide favored this quest, and with a tidal range of 30 feet we could be sure that any shoal covered at this tide would present no obstacle to our ships at top flood.

A few minutes' flight to the southeast brought us to Bartlett Narrows, a distance that the *Morrissey* took five hours to traverse. We circled in order to examine the waters at the entrance to the Narrows and to photograph the area from various bearings. I now saw for the first time a rock exposed by the low tide that had been submerged every time we had been there in the *Morrissey,* and two others near enough the surface to be visible from the air, and near enough the channel to menace navigation. One was just off the point of Low Island and uncomfortably close to the range that Crowell had selected for guidance through the passage. With pencil and paper I sketched the channel and made careful mental notes for the task of piloting the *Bear* down to the lower bay. A slightly zigzag course looked safest, first to right, then to left of Crowell's range.

From there we flew on down the bay another 35 miles to examine what the chart showed as a large island some 11 miles long, labeled Chase Island. When we got there we found it to be in reality an archipelago of a dozen or more islands, large and small. A few photographs sufficed to record this, and we returned to Koojesse Inlet after a most profitable hour and a half in the air.

I learned on our return that Colonel Ghelard, who was inspecting the base, was to leave the bay on the *Bear* that evening when the inspection was completed. The responsibility of piloting the ship through the still uncharted Bartlett Narrows was something I should have preferred to tackle in full daylight on a rising tide, but the skipper of the *Bear* was in a hurry to get on his way and couldn't wait till morning. I asked him if he was willing to go through at night on a falling tide. He was, so off we went. I was a bit apprehensive about threading our way in the twilight among the rocks I had seen from the air, but the sky was very clear, a full moon was rising, and the *Bear* was equipped with good binoculars and a reliable fathometer.

Down the bay we went as the twilight deepened and the moon rose higher. We approached the Pink Lady, and I consulted the little sketch of the channel, hastily drawn in the airplane that afternoon; it was the only chart we had. Commander Graves called to me from the bridge, "Come out and look at the moon."

I stepped out of the pilothouse and looked. There was an ominous curved bite taken out of the starboard side of the moon's disc. I had completely forgotten the total eclipse scheduled for this night—of all nights for the moon to put on such a show! The earth's shadow advanced rapidly and the moonlight waned, but soon I picked up with the binoculars the clear outlines of the Narrows and the steep mountain beyond. Eclipse or no, all I had to do was to swing a bit to starboard of the range till within half a mile of Low Island, then cross the range to port, and then aim for the Narrows. The moon shrank to a slender crescent. The fathometer registered rapid shoaling, but we crossed these submarine ridges without getting near enough the bottom to disturb the skipper's equanimity.

At last we rounded the point of Low Island and entered Bartlett Narrows. Just at that moment the eclipse of the moon became total. The aurora was so brilliant that it fairly took the place of the moon to light us on our way; indeed the moon by contrast looked like a dark spot in the sky. But our worries were really over, for the shoals were already behind us, and the bold shores of the Narrows were clearly visible as we sped through and out into the wide, deep waters of the lower bay.

Breathing more freely, we watched the receding shores of the island barrier to starboard in the bright glow of the aurora, and laid our course southeastward down the bay. The rim of the moon reappeared, sun-illumined, and soon the bay was bathed in bright moonlight, casting a silver wake where the waters were ruffled by a gentle breeze. Beyond the wake lay the dark outlines of islands, while farther still rose the blue mountains of the southwest shore of the bay. Rarely have human eyes feasted on a lovelier picture.

So accustomed had I become to the slow pace of the *Morrissey* that I was surprised when the skipper called my attention to a fiord on the starboard beam that he thought might be Kneeland Bay, where we were to meet Captain Bob and the sturdy schooner. I looked through the powerful binoculars, and sure enough, there in the center of the fiord was the sharp peak of Anvil Mountain, an unmistakable landmark. We turned and headed straight for the mouth of the fiord.

The *Morrissey's* fathometer had shown 100 fathoms in the

middle of Kneeland Bay and nothing less than 30 at the entrance. I therefore assured the skipper that the mouth of the fiord was wide and deep and free of hazards. Quickly we approached the fiord, heading for the middle of the entrance, when to my uttter amazement the fathometer showed rapid shoaling of the water. The skipper rang "slow" on the engines, but the bottom continued to approach the keel at such an alarming rate that he rang "stop" and then "full astern." Before the ship lost all headway the fathometer reading decreased to zero—no measurable depth of water under the keel! In a few seconds our slight remaining headway carried us on into deepening water, and the skipper, quite unruffled, resumed progress at a cautious speed. I could scarcely believe my senses and wondered if it was all a bad dream, for in our two previous visits to this fiord nothing had suggested a shoal at the entrance. The mouth of the fiord was more than a mile wide, and the *Morrissey* had always followed the shore at one side or other of the entrance; never had she taken this middle course followed by the *Bear*. During the few seconds that we were passing slowly from the shoal to the deep water beyond, I noted such ranges as I could see in the moonlight, to guide me in finding and sounding the area by daylight. Later we named it "Nightmare Shoal" and when the opportunity came planted an oil drum there for a buoy. All around it the bottom fell away to depths of 100 fathoms and more. Captain Bob was greatly interested in this strange submarine plateau. "They'd have crucified you if you had put the *Bear* on that!" he exclaimed with forceful emphasis.

The fathometer gave us no further alarm as we went on into the fiord in search of the *Morrissey,* but the skipper preferred not to follow her into her anchorage behind the island, under the deep shadows of the great cliffs to the south. He stopped instead well out in the middle of the fiord and sent the tender in to search for the schooner, and a search it was, for the shadows were dark under the beetling crags. Assuring the skipper that he would find deep water along the south shore of the entrance, I bade him good-bye and boarded the *Morrissey* at 2:45 A.M., quite ready for a sound sleep. The *Bear* slipped out of the fiord, down the bay and out to sea to take Colonel Ghelard on his next errand for the Air Corps.

12 The New Air Base

On August 29th the *Morrissey* returned to Koojesse Inlet in a stiff southeaster that swept into the harbor. By nightfall it was too rough to permit the usual transfer of the airport construction gang to the *Fairfax,* laying at anchor outside the harbor. Some two hundred workmen were thus marooned ashore for the night. The shacks and tents already erected were not nearly enough to house them all, even lying packed like sardines on the hard board floors. Many of them lay on the ground outside, huddled under big tarpaulins and keeping each other warm as best they could.

The gale raged through the next day, making it impossible to do any sounding, but it did give me time to catch up on placing the previous soundings in their place on the harbor chart.

Building an airport in such a place as this, without any dock for the landing of supplies, calls for resourcefulness and judgments not required in most places. The tug *Mathews* brought in two large barges for unloading the vast store of trucks, bulldozers, and machinery of all sorts, and ferrying them ashore to the airport site. As there was no time to build a dock, a lane on the beach was cleared of boulders from high-water mark far enough down to afford a smooth area of firm sand on which the barges could be grounded at high water and unloaded when the tide went down. The ships had come in just before the full moon and the contractor's gang thus enjoyed high spring tides for the first five or six days of their task of lighterage. With the maximum tidal range of 38 feet, the distance along this beach uncovered by the falling tide is about 1,000 yards. As the moon waned and the tidal range grew less, Crowell sounded a warning that on this nearly level beach each flood tide would fail by many yards to reach the mark of its predecessor. But the contractor's men could not believe that their range would decrease so rapidly, so a day came when they found their barge still hard aground when high tide came and went. For nearly two weeks, while the neap tides prevailed, this barge lay stranded and the unloading had to be done entirely by the other.

With a vast number of landlubbers ashore and afloat the inlet turned into an aquatic madhouse. The turmoil of traffic in this hitherto wild and lonely harbor was unbelievable. Motorboats whizzed to and fro between the steamers and the beach, carelessly steered and barely missing the reefs that we had surveyed. This got terribly on Captain Bob's nerves as he watched from the deck of his beloved *Morrissey*. It was a relief to him and to all on board when we decided to go down the bay and commence a proper survey of Bartlett Narrows, the new passage of our choice for bringing ships through the island barrier.

With fathometer slung from the *Morrissey's* rail, we recorded a good profile of the bottom as we crossed the bay, and a striking profile it was. Just as from the air one sees the islands in Frobisher Bay arranged in straight rows aligned northwest and southeast, parallel to the long axis of the bay, so the fathometer revealed a succession of deep troughs, similarly aligned, and the transverse profile is a rapid succession of steep ups and downs.

As we approached Pink Lady Island, the Captain suggested exploring the waters east of it, which hitherto we had avoided. We were rewarded by finding a wide, deep channel close to the east shore of the island, leading more directly toward the base at Koojesse Inlet than the route used hitherto. This new channel at once became the regular route. Well pleased with this find, we sounded our way into an anchorage by the south side of an island

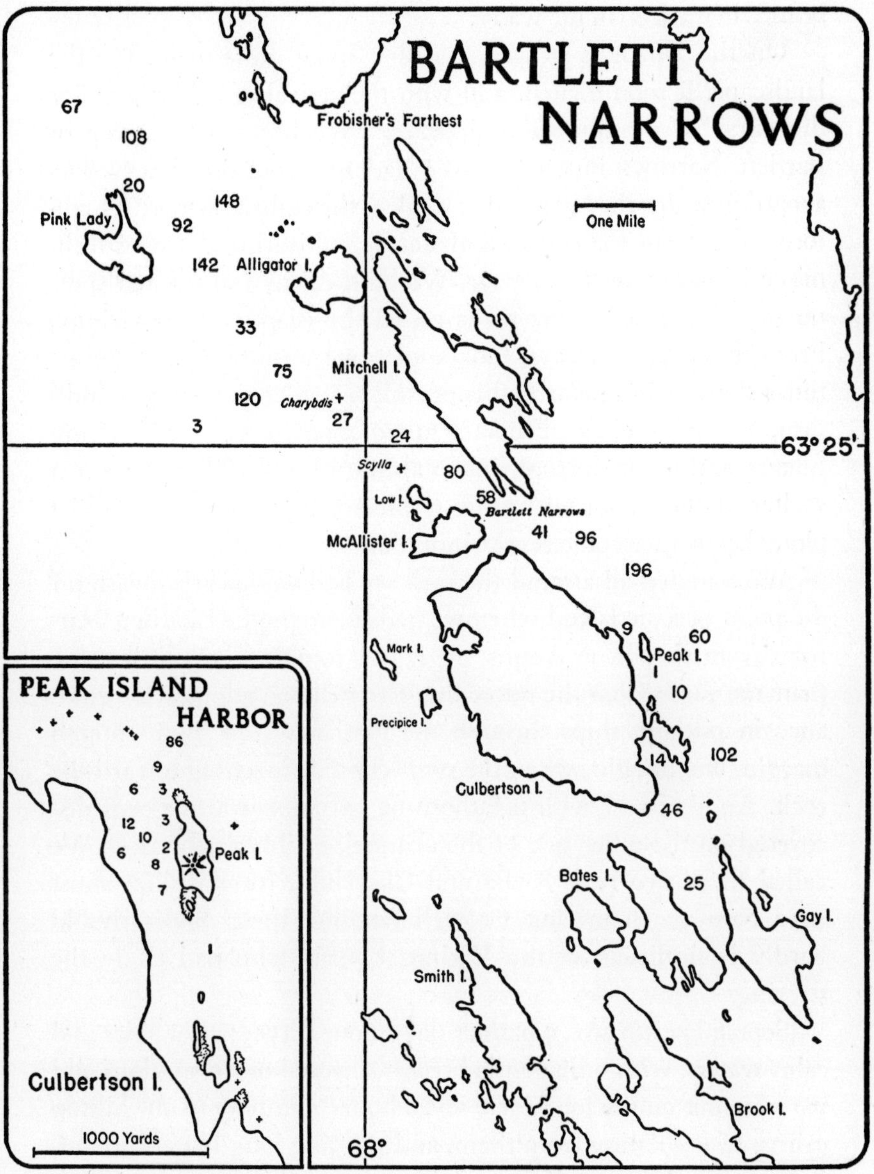
BARTLETT NARROWS
One Mile
Frobisher's Farthest
67
108
20
Pink Lady
148
92
142
Alligator I.
33
75
Mitchell I.
120
Charybdis
27
3
24
63° 25'
Scylla
80
Low I.
58
Bartlett Narrows
McAllister I.
41
96
196
9
60
Mark I.
Peak I.
1
10
Precipice I.
14
102
Culbertson I.
46
Bates I.
25
Gay I.
Smith I.
Brook I.
68°
PEAK ISLAND HARBOR
86
9
6
3
3
12
10
2
6
8
Peak I.
7
1
0
Culbertson I.
1000 Yards

we had named "The Alligator" because of its resemblance to that beast when asleep in the water.

On the 3rd of September the first snow came to whiten the landscape. It soon melted, and with the arrival of colder weather the curse of mosquitoes disappeared. We began our survey of Bartlett Narrows in earnest. We had no professional surveyors aboard now, for they were all at work at the embryo airport. Therefore we got out the transit and started cultivating our own technique in mapping the passage. We built cairns and painted spots on the steep faces of the rocks on all the islands in the vicinity. From these landmarks we took angles by transit or sextant. Sometimes the weather favored this pursuit; at other times cold winds, sleet, and snow clouded sextant mirrors, nearly paralyzed aching fingers, and made the soggy notes almost illegible. Then the warm shelter of the cabin, where the results already achieved could be plotted, was a welcome respite indeed.

We sounded all around the rock we had seen nearly awash off the point of Low Island when we had photographed Bartlett Narrows from the air on August 25. It was found to be a little more than 100 yards from the range that Crowell had adopted for guidance in piloting ships through the Narrows—not quite enough margin, we thought, when the swift ebb tide is setting toward the rock. Another reef with a fathom or two at low water was discovered on the other side of the channel, off Mitchell Island. We called these two reefs Scylla and Charybdis, for the ships must steer between them; but Crowell felt that these names should hardly be inflicted on the Viking skippers who had to do the steering.

September 5th was a banner day in our survey; clear skies and calm waters, with only gentle breezes, favored our operations and we were out on the job nearly seven hours, climbing to our hilltop cairns, triangulating from them, and running long lines of soundings in the whaleboat. The next day was even better, with mild air that made it a joy to be afloat. We tried to buoy both Scylla and

Charybdis. The black buoy on Scylla was a success, showing at most tides, but towing under when the current ran its hardest. With Charybdis we had no luck; the anchor must have slid down the steep slope of the rock into deep water, for the buoy was quickly dragged under and disappeared.

By the afternoon of the 6th we had explored Bartlett Narrows enough to know its great depth and how to approach it safely, and took advantage of the fine weather to investigate the harbor behind Peak Island. Sounding with the fathometer we found between the islands a nearly level floor about 10 fathoms down, allowing ample room for two or three ships the size of the *Morrissey* to swing at anchor. The entrance was narrow, but wide enough to admit a bigger ship than the *Morrissey,* and deep enough for any ship. This discovery was to prove of great value later on.

That evening a long radio dispatch was received on the *Morrissey,* including a message that I was to deliver at once to the captain of the *Fairfax* telling him to proceed to Boston as soon as his ship was unloaded and directing me to arrange pilotage through the islands. All this was backed by urgent orders from General Brehon B. Somervell. Although two days more in Bartlett Narrows would virtually have completed our local survey, the message had to be delivered.

Early in the morning we were under way for Koojesse Inlet, pausing only to anchor an oil drum for a buoy by Charybdis. This time we used rope, for the buoy must be anchored in 15 fathoms, and the requisite length of chain would have dragged it under water. The trick worked and the buoy remained to mark the channel.

We arrived at the Inlet soon after noon, delivered the message to Captain Danielson of the *Fairfax,* and found Crowell in the midst of a perplexing dilemma. A cargo ship, the *Denny,* was due to arrive in the bay on September 10 and was to be met near Kneeland Bay. Crowell's job included taking the *Fairfax* out and bringing the *Denny* in. When would the *Fairfax* be unloaded and ready to go? And if at the same time as the *Denny's* arrival, which should he do first? We were told that the *Fairfax* would probably be ready to sail on the 9th and on that day Crowell should start, if he was to meet the *Denny* 60 miles down the bay on the 10th. Two days remained in which to work out the program, and this gave me time to look over the fast growing base and note the vast changes that had already taken place in the fortnight since the steamers had first entered the bay.

I wandered through the enormous tent colony and saw the commissary store in operation—where one could really buy things as in a frontier department store. Prefabricated houses for the men and a great warehouse for stores were already under construction, power lines were strung, generators poured their currents through them into hut and messhall. Great mechanical dinosaurs walked the streets at the behest of the contractor's genius, while far out on the runway bulldozers were pushing tons of dirt about.

On the morning of September 10 we started down the bay to meet the *Denny,* leaving the *Fairfax* at the base to await our return. On the way we stopped at Crowell's camp to pick up two machine guns that Crowell had taken there in a hurry by motorboat from the *Fairfax* a week ago. I now learned the meaning behind some of Captain Bob's cryptic and jesting remarks about "Tony's sub." Tony, the sled-dog man at Crowell's Island, thought he had seen a periscope in the harbor. A real blackout had been put on and Crowell had hastened to arm his camp with the machine guns. Now it was presumed that Tony had seen a seal raise his curious head above the water, and the machine guns were to be returned to the *Fairfax.*

On September 11 we had heard nothing from the *Denny* and radio messages from the *Bear* urgently instructed us to return for the *Fairfax.* Captain Bob received a message relaying orders from Colonel Dunn of the Corps of Engineers to the effect that as soon as all survey work and delivery of charts were completed he would be relieved by the *Amarok* and released from any further duty in the area. Joining a southbound Canadian convoy he was to proceed to Boston. Here then was a welcome intimation that I should soon be bound for home.

We were under way at 4:30 A.M. the next morning and arrived at Koojesse Inlet at 2:40 P.M. Crowell made preparations for his forthcoming trip, for the schedule was a tight one that required a great deal of planning. Taking the *Fairfax* out, he would meet the *Amarok,* bring her in, then take out the *Eleanor*. Still there was no further news of the *Denny;* we began to wonder if such a ship really existed.

On the morning of September 13 we saw the *Fairfax* move out of Koojesse Inlet on the dot of 5 A.M. and shortly thereafter we mobilized the whaleboat for a final sounding tour of the harbor. For the first time we made a complete circuit of Long Island and covered the hitherto unsounded areas thoroughly enough to make our harbor chart adequate for all practical purposes.

That evening we saw a ship approaching. It proved to be the long-lost *Denny* with Crowell aboard. He spent the night aboard the *Morrissey* and early the next morning was off again to take the *Eleanor* down the bay and bring back the *Amarok,* which was waiting for him in the Lion's Den, the harbor at Daniel Island.

The afternoon of September 14 was a fine one with a clear northwest wind, and a respite from chart construction seemed in order. Rigby and I went ashore together for a sight-seeing ramble over the hills. A band of Eskimos had recently come up from somewhere down the bay and made camp on the lovely patch of green grass by the rapids of the Sylvia Grinnell River. We walked across the broad flats, then over the high ridge and down to their camp, where we were highly intrigued by their primitive ingenuity and picturesque costume. Thence we returned to the base, where Rigby had to visit the newly erected hospital (a Nissen hut) and begin a strenuous two days of seeing a steady stream of patients. With them as clinical material, he gave what he could of a four-year course in medicine to the "first-aid man," who hadn't even had a first-aid course, his only equipment being some textbooks and a fairly complete line of medical supplies. Here was a heavy load for a novice. The *Fairfax* had gone, taking her doctor with her. Cargo ships came, unloaded, and hurried away. Now Rigby was leaving in a few days, and the army of airport builders would be left at their hazardous task without any doctor at the base. Dr. Joachim at Crowell's camp 30 miles away could be brought by motorboat at about eight hours' notice, in case of emergency. But without a doctor at the base, the possibility of trouble looked grim.

While Rigby was running his clinic, I picked up a textbook of physiology from the dispensary shelf. There I found a reference to some work I had done in the Harvard laboratory so long ago that in retrospect it seemed like an earlier incarnation. So I had once been a physiologist, had I?

Rigby had brought back aboard the *Morrissey* some strange tales of life aboard the *Fairfax*. The men were getting big wartime wages—$500 or more a month in some cases; but there was little for them to buy with their new-found wealth. One evening in the main saloon there was a poker game; one workman won $800 without rising from his seat, decided to quit with that; left the table, then decided to have another go at it, sat down and won $300 more, thus cleaning up $1100 in a single evening. What to do with the vast roll of bills perplexed him. Someone suggested a present to the visiting doctor. There was ginger ale to spare aboard the *Fairfax,* and Rigby, accepting a case of it as a good-will gift, brought it aboard the *Morrissey*. How the balance of the winnings was blown in, history does not relate.

We heard another tale of fabulous figures from the camp. An elbow joint, such as could be bought for 15 cents in a hardware store, was needed for a stovepipe being installed in a tent. As the camp supply was all out of such things, the metal-smiths had to make one out of their sheet metal supply; the charge for their labor came to $22.

On the morning of September 15, the *Amarok* arrived at Koojesse Inlet, piloted by Crowell. She turned out to be none other than the *Lark*—Grening's trawler flagship in our combined Crystal voyage of October 1941, now purchased by the U. S. Navy and renamed. Her skipper, Lieutenant Crouch, came aboard the *Morrissey* to begin the business of relieving us of our duties. The *Amarok* was to be guard ship and handle communications and pilotage after our departure. Crowell had brought her in by Cincinnati Press Channel and introduced Crouch to his island base. I then showed him what else we had learned of the geography of Frobisher Bay, and in particular Koojesse Inlet and the approach to it through Bartlett Narrows. More detailed instruction on the narrows was to follow on the spot, when we could look over that passage together.

On the 17th the *Morrissey* made her seventh and last departure from Koojesse Inlet. We had arranged to meet Crouch the next day at Bartlett Narrows and acquaint him with his new duties.

Captain Bob was interested in the idea of using the harbor behind Peak Island for an anchorage; the shelter within tempted him greatly, but the entrance was narrow, and the ebb tide swept across it in a way that threatened to put him on the point of Peak Island. I sounded the entrance in the whaleboat and reported my findings. The Captain hesitated a bit, but then the spirit of exploration got the better of him. In went the *Morrissey,* with a margin of a good fathom under her keel. Once inside we anchored in the snuggest shelter we had yet found in all Frobisher Bay; the skipper was well pleased.

Next day the *Amarok* came up the bay escorting the *Norluna,* a big ship bound for Koojesse Inlet with Crowell aboard for pilot. We intercepted the *Amarok* with the whaleboat and led her into our new haven of refuge, where she anchored beside the *Morrissey.*

Setting out with Crouch in the whaleboat, we killed two big birds with one stone: we made a long sounding run through the narrows, zigzagging as necessary to fill in any remaining gaps in our survey, while I gave him complete instruction in pilotage through the passage.

Returning to the *Morrissey,* I set to work completing the reconnaissance charts to be delivered to Lieutenant Crouch on our departure. They included a general chart from Kneeland Bay to the base and five large-scale charts showing the Lion's Den, Koojesse Inlet, and the passages leading thereto. Rigby and Majer turned to and helped and by 3:00 A.M. the charts were all drawn and copied so that Crouch could have one complete set and I could take another back to Washington. My job was done.

The morning of September 20 found us homeward bound, exactly two months after our first entrance into Frobisher Bay eager in our anticipation of the adventures that lay before us. We had not been disappointed. By airplane, schooner, whaleboat, and fathometer we had learned much of its geography. The majesty of its mountains and the charm of its islands thrilled us; and the drama of seeing great ships enter, start the building of a great airport, and go their way to carry on their work in the war had filled the days with memorable incidents.

Reports had reached us of enemy submarines sighted in Hamilton Inlet. As we started across that sheet of water, I began to inform the radio operator of our precise position every hour, in case we should sight a sub and thus be in a position to give our bombing planes an important clue. Captain Bob scoffed at the rumors of subs in Hamilton Inlet, likening them to "Tony's sub" in Frobisher Bay. "It's in the Strait of Belle Isle that they are working," he said in a marked manner. They had indeed been sinking schooners there; and as we approached and crossed the Strait the skipper seemed to me more tense than usual. Belle Isle and then Cape Bauld appeared ahead and disappeared astern, and after that my plans for the quick dumping of our cryptographic gear over the side no longer exercised my thoughts.

On the 2nd of October we sailed into Brigus and tied up to the wharf. Captain Bob then set foot ashore for the first time since

July 10. Not once in Port Burwell, Frobisher Bay, or Labador had he touched dry land. He took me to dinner at his family home; then, leaving the ship at Brigus, I made my way to Argentia, and on October 6th flew in a Navy transport plane home to Boston.

PART THREE

- 1943 -

13 Mapping and Planning

The winter of 1942-43 was one of momentous development in the airplane ferry service. Immense numbers of planes, all but the short-range fighters, were being flown to Europe over a variety of routes. On the great terrace by Goose Bay, where in July 1941 we had seen nothing but a level carpet of caribou moss dotted with slender spruce trees, the Canadians had three runways in operation within five months. Then in 1942 the U. S. Army Air Corps expanded the field into one of the greatest airports in the world, with enormous hard-surfaced runways and a fast-growing collection of hangars and barracks. At Narsarssuak near the southern cape of Greenland a large airport was built at the head of a fiord. These two fields served as steppingstones by which all but the planes of most limited range could be flown from Presque Isle, Maine, to Iceland and thence on to Scotland. By the summer of 1943 many planes were being sent over this route to the combat zone.

Farther north, at the head of Sondre Stromfiord on the west coast of Greenland, another airport was built, intended primarily as a link between our Baffin Island fields and Angmagsalik on Greenland's east coast in the same latitude. As things turned out, it was used primarily as an alternate airport where planes could go when the weather was too bad at Narsarssuak.

Throughout the winter, work was pushed vigorously on the airports at Chimo and Frobisher Bay, to complete the chain for short-range planes over the northernmost route, via Sondre Stromfiord. Cargo ships continued to deliver supplies to both bases, and the last ship to leave Koojesse Inlet in the fall had to break her way through 5 inches of ice to reach the open sea.

From Boston I had returned to Washington and resumed work in the Hydrographic Office, now established in its large new building in Suitland, Maryland. My main task was to make the best charts of Frobisher Bay that I could from the material thus far gathered. This consisted of Hall's map of 1865—still the basis of existing charts; the record of four geographical positions carefully placed by Canadian surveyors; one position at the head of York Sound, fixed and marked by trained observers on Miss Boyd's expedition of 1941; a large but incomplete set of oblique photographs taken from the *Bear's* plane on two flights in August; vertical photographs of Koojesse Inlet and Cape Rammelsberg, taken on Elliott Roosevelt's flight in the sumer of 1941; and finally, my local triangulations and soundings of three significant areas, and miscellaneous notes collected in various parts of the bay.

We now had large portions of Frobisher Bay covered by oblique photographs, but only two small areas shown in vertical views. Therefore a big job of oblique photogrammetry lay before me. My wife, with a skilled hand trained in art schools, was eager to help, if she could, in drawing the charts. The head of my division in the Hydrographic Office encouraged her to come as a volunteer and do a share of the drafting, but the head clerk told us that to work for the Government as a volunteer without compensation would mean a jail sentence; such is the inscrutable law of the land. After tedious days of red tape, she received a Civil Service designation that made her friends laugh, and went to work with me at a desk in Suitland, learning to rectify oblique photographs.

Photogrammetrists define obliques as photographs taken with the camera purposely aimed in a direction between horizontal and vertical, and classify them in two categories, high and low. High obliques are those in which the horizon is shown. In low obliques the horizon is not shown, the camera being usually aimed more nearly in a vertical than a horizontal direction. The most elementary way to map a shoreline from an oblique photograph is the Canadian grid method. A perspective grid, representing squares on the ground, is drawn on transparent material such as frosted acetate, to be superposed on the photograph. The squares are drawn as they would appear in the oblique photographs, their sides receding into the distance shown as lines converging to a vanishing point on the horizon. Properly adjusted to the photograph, these obliquely viewed squares show just where the shoreline intersects the sides of the corresponding true squares in a properly drawn map. True squares are therefore ruled on the map paper, and by visual comparison with the photograph a rectified map is drawn. The perspective grid and its use in mapping two sample islands in Frobisher Bay are shown in the accompanying diagram.

As we labored with this rectification of many photographs, my thoughts turned to my friend Miller and his single-eyepiece plotter at the American Geographical Society, the instrument with which he had rectified the Labrador shoreline, and which I had used in 1941 to map Whitefish Lake near Chimo. If only we could have such an instrument at Suitland we could draw the shoreline of Frobisher Bay more quickly and accurately than with the Canadian grids. Oblique photographs not only cover more ground than verticals, but in wartime they are easier to take in a zone of military operations, for when a plane flies directly over hostile territory, either intercepter planes or antiaircraft gunfire may interfere and bring the photographic mission to a quick and tragic end. Before the war could be over, many oblique photographs of many important places might need speedy rectifying. With this prospect, Miller was consulted. His plotting instrument, after long use, had developed mechanical defects, but he had a plan in mind for a greatly improved model. A request to the Coördinator of Research and Development in the Navy Department resulted in a directive to the National Defense Research Committee to investigate and develop instruments and methods for plotting maps from oblique aerial photographs. The actual development of this project was to be carried on by Merrill Flood and Associates of Princeton, New Jersey.

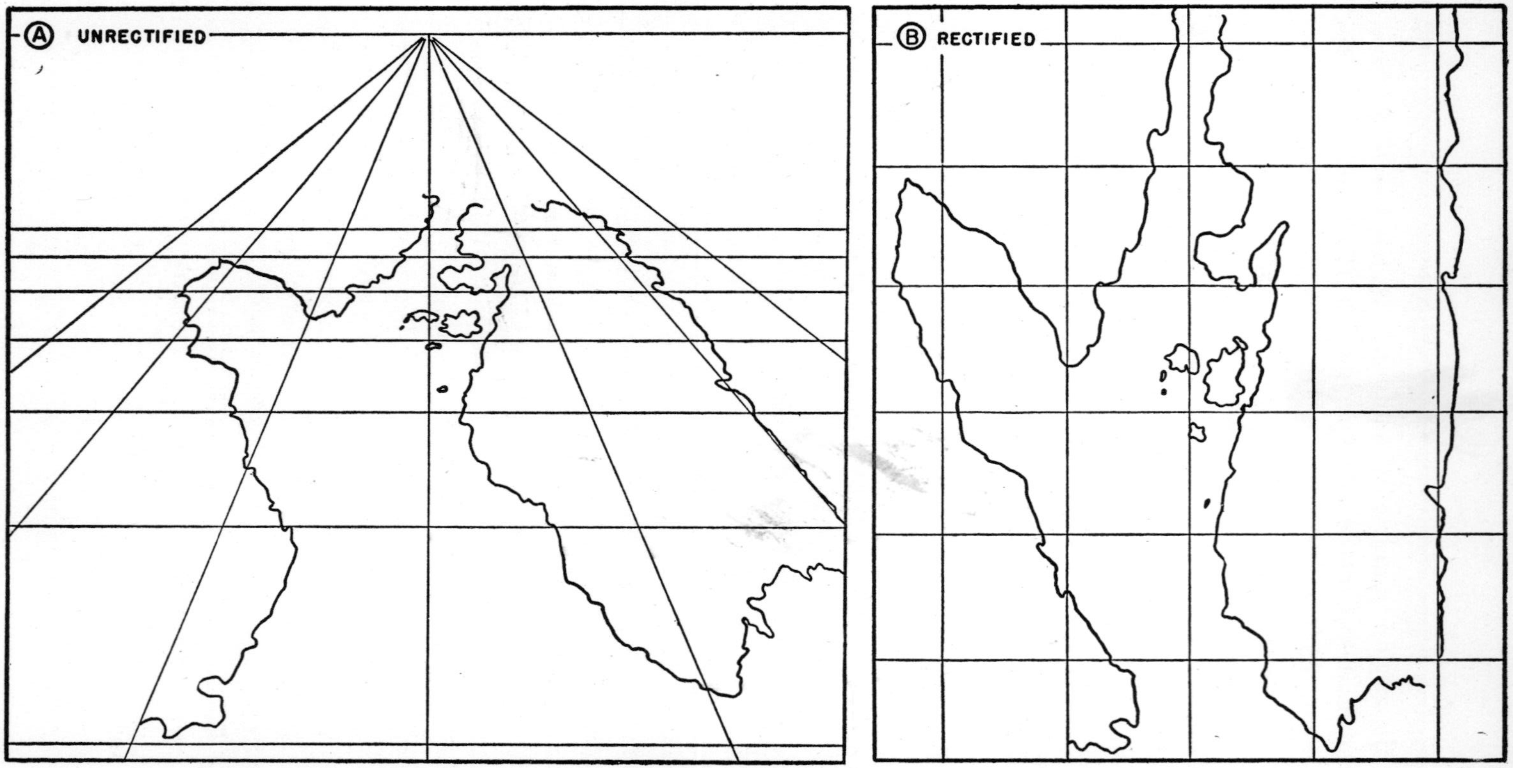

Shoreline mapped by Canadian perspective grid method: (a) tracing from oblique photograph; (b) rectified outline.

My assignment as liaison man for this project and construction of the Frobisher Bay charts kept me busy through the winter. Hall's ancient map still furnished the basis of our chart for those parts of the bay on which we had no new information; but our oblique photographs enabled us to draw many miles of shoreline with a precision far beyond what Hall's facilities had made possible.

By April 1943, it had been decided that the Navy should send parties of experienced hydrographic surveyors to the Koksoak River and Frobisher Bay, to do proper surveys of both areas, and a directive to that effect was sent by the Commander in Chief to the Hydrographic Office. It is quite out of the ordinary to put a medical officer in charge of a hydrographic survey, but since I had seen the Koksoak River both from the air and from the bridge of a trawler and had spent two months on reconnaissance in Frobisher Bay, it fell to my lot to be designated officer-in-charge of the combined project. Officers specially trained in hydrographic work were to direct the technical work in both areas; thus my job was that of local guide, rather than technical expert.

Three officers of experience as hydrographic surveyors were selected for the expedition: Lieutenant J. Stirton, Lieutenant (j.g.) E. Mohl, and Ensign T. Gunther. Two other officers in the Hydrographic Office with previous experience in geology and land surveying were also assigned to the project: Lieutenant (j.g.) S. A. Wengerd and Ensign D. S. Turner. Some enlisted men with special training in "Hydro" and others for less expert assistance were also allotted to us.

The two areas to be surveyed were very different. The Koksoak River, about a mile wide, flows north 30 miles from the base to the sea. At the river mouth an area of about 8 by 10 miles must be dealt with to cover the shoals that ships had to dodge in finding the entrance. In contrast to this, Frobisher Bay, 170 miles long and in places 45 miles wide, has many times the area of the Crystal One approach, but its waters are deep, and the safe pilotage of ships to the base would require detailed soundings only in a few small areas, already fairly well mapped in the reconnaissance of 1942. The Koksoak River, though limited in extent, was a passage of unusual danger to ships, for though the river is a mile wide, the navigable channel is less than half of that and is bounded by enormous boulder trains concealed at high tide by the muddy water, which rushes past them on the ebb tide at an 8-knot speed. The dangers of this river were brought home with tragic emphasis in the late fall of 1942 when a tugboat with a barge in tow entered the river on a flood tide. The tugboat skipper, unable to summon the guardship from the base in time to pilot him up the river before dark, and fearing the stormy weather outside in Ungava Bay, attempted to find his way up the river unaided. Guessing at the channel, he guessed wrong. The strong flood tide swept tug and barge broadside against a boulder train. The tug was wrecked and the skipper and most of his crew were lost. A few survivors reached the shore, swimming in the icy water, and shivering through the wintry night, barely missed dying of exposure.

The Koksoak River is ice-free from mid-June to November, whereas the head of Frobisher Bay is locked in solid ice through July, and may begin to freeze again early in October. Now that airports were open for business in both places, the transport of a survey party was much easier than previously. The plan was to send us all to Crystal One at Chimo by airplane in June to start work on the Koksoak River. The *Morrissey,* loaned by the Army to the Navy, would come to Chimo as soon as she could get through the ice in Ungava Bay, bringing some of the heavy supplies needed for the survey. Then Wengerd, Mohl, Turner, and I would sail on the *Morrissey* to Frobisher Bay, leaving Stirton and Gunther to finish the survey at Crystal One. Big transport planes were already delivering mail and fresh supplies to both bases, landing on hard surfaced runways, and could bring the enlisted men to either base at short notice.

Enough small craft were reported available on the Koksoak

River to serve for our operations there, but we needed a sounding boat for the work in Frobisher Bay. In the Navy's Bureau of Ships in Washington was a lady whose mind contained a veritable catalogue of all the small motorboats acquired by the Bureau. Arrangements were made to have a 32-foot cabin cruiser delivered in Boston, where she could be prepared for the rigors of Baffin Island and sent thither on a cargo ship when the ice should loosen its grip on the bay.

14 Surveying the Koksoak River

The afternoon of June 12, 1943, found us aboard a plane winging northward towards the Koksoak River. We landed on a long runway diverging but slightly from the imaginary landing strip that I had paced from the river bank in July 1941.

Snow flurries and icy winds greeted us here in this northern outpost, close to Ungava Bay, but from my previous visit I knew we could expect warm weather soon. Blocks of winter ice still floated down the river, but not in such numbers as to bother the small boats in which we should soon be plying its waters.

Soon after our arrival I met Freddie Edmunds, my guide to Whitefish Lake on the memorable ramble in October 1941. Now he was thirteen and already earning his keep by helping his father as interpreter in the liaison between the construction gang and the friendly Eskimos. He greeted me cordially with his endearing smile, as an old friend. When next I met his father, I spoke with pleasure of having seen Freddie again.

"Yes," he replied, "Freddie said he had seen his grandfather, and I didn't know who he meant." This remark I should hardly dare record if I could not prove a perfect alibi.

The building of such a base in so inaccessible a spot was a noble achievement. All heavy gear must be brought in ships during the brief season when Ungava Bay is free of ice. The ships must be met and piloted up the dangerous river where, till this year, none but Edmunds and his group of Eskimo pilots knew the narrow channel between formidable boulder trains on either side. From the anchorage by Chimo all gear must be lightered up to the base on shoal-draft barges operating only on the high tide when the boulders on the shoals were submerged to a safe depth.

Combatting these fearful odds, the contractor and the Army had built a livable base with long runways on which big cargo planes could land with fresh supplies and mail throughout the year. Army doctors had equipped a small dispensary and hospital with the needful things to meet almost any emergency, and considering what they had to work with, they had done a job of which they could well be proud. On June 15 the new movie theater was completed and formally opened, and the next day a new enlisted men's mess hall was inaugurated in the room hitherto used as a movie theater.

That evening after supper I was offered an opportunity to fly in a DC3 to Frobisher Bay, sleep in the now well-appointed Crystal Two base and return next day in time for an afternoon's work with my gang. It was too good a chance to lose. Flying north over the ice floes of Ungava Bay, we steered straight toward a flaming sunset; we were near the Arctic Circle and getting nearer every minute. The red sun seemed to roll slowly along the horizon toward the North Pole. Massive clouds hid Meta Incognita, but we broke through them as we reached the shores of Frobisher Bay, swooped down over Koojesse Inlet, and landed on the great new runway just at midnight. In this latitude it was almost full daylight, and the moon in the southern sky contributed more decoration than light.

Next morning as we started back to Crystal One, I examined the upper bay. It was frozen solid, and although it was now June

17, not a crack in the ice could be seen all the way to the island barrier. Flying over the mountains, we came to the south shore of Baffin Island whence open water extended southward about twenty miles till the pack ice began abruptly and extended without a break to the shores of Ungava Bay. We came down to within 400 feet of the surface—you can do that safely in a twin-motor Douglas when the weather is fine—and had an intimate but fleeting view of the details of the ice pack. Once or twice we surprised a seal basking in the sun by a narrow open lead and saw him slip into the water as we passed close over him.

When still far out over Ungava Bay many miles from Cape Hopes Advance, we came on a party of Eskimos with their komatik and dog team, hunting seals on the ice floe. They waved to us and we circled low to get a good look. In the middle of a large bloodstained patch of ice lay the seal, while men and dogs rested after the kill. The Eskimos hanging round the Hudson's Bay Posts and Moravian Missions were no longer a glamorous novelty to me; but this fleeting glimpse of them at home on an ice floe far out at sea, leading the nomadic life of their forefathers, brought back vividly the pictures I had seen in books on Arctic exploration, pictures that stirred a wild desire in my childhood to know the far North in all its austerity. This was the real thing.

Back at Crystal One preparations for the survey were progressing well. A motor scow was fitted out for the aquatic work of the survey and duly christened *River Rogue*. The next week six of us set out in her and ran down through the islands to begin the erection of tripods and shore signals for the survey. The procedure was to set up a "network" of primary stations—tripods built on high points from which they could be seen a long way off, and from which the shores of the river would be visible. These were to be accurately triangulated and thus would become the skeleton of the survey. Then secondary stations were to be placed near the shores, usually in the form of "twisters" with something like a squaresail rotated at the bottom to make it visible from all sides. These were to be triangulated from the primaries and serve in turn as landmarks by which to place the soundings when the hydrographic work was done.

Pleasant days ensued on the river, all hands journeying downstream in the *River Rogue,* dividing into squads to put up the tripods and twisters at suitable headlands. Our path brought us daily past a tragic sight: the cargo ship *Norluna* which had visited Frobisher Bay the previous summer had met a watery death in the Koksoak River and now lay derelict on its shore. Late in the fall she had come to Fort Chimo with supplies for Crystal One. At dusk a sudden gale from the north started the ships dragging their anchors. The *Norluna* had no steam up and her anchors dragged till she was driven hard aground on the north point of Little Elbow Island. There she stood, bolt upright, her starboard side nearly touching the precipitous cliffs, and the rocks on which she rested fitting into her battered hull as neatly as a dentist's inlay fits the remnants of a tooth. Her valuable cargo of oil in barrels was buried deep under a solid mass of ice, for the river water flowed freely through the gaps rent in her plating, and the warmth of summer had hardly begun to melt it. One day we went aboard and explored the abandoned ship; we found chaos and order alternately prevailing in different parts of the vessel. A number of cans of tinned pemmican were temptingly displayed. We sampled one and found it excellent. Later we used the ship's rudder post as a tide gauge, wiring a calibrated rod to it so that the tide level could easily be read from the shore with binoculars. Late in the summer much of her oil cargo was salvaged.

On June 26 Wengerd arrived with seven more enlisted Navy men. Thus augmented, our force pushed the work ahead faster. As the net of tripods extended more and more miles down the river toward the sea, triangulation with transits began, following closely the progressive erection of new tripods. By July 6 enough shore signals had been located for us to draw up a boat sheet and start sounding the first few miles of the river, nearest the base.

An Army M-T boat was placed at our disposal for the construction gang to put up tripods and twisters ever farther downstream, and the *River Rogue* became the sounding boat, with Gunther in charge.

On July 9 Stirton arrived. He was our most experienced hydrographic expert and was to take full charge of the Koksoak River survey when the time came for the Frobisher Bay contingent to proceed north. The network of tripods was now extended so far down the river that a prohibitive amount of time was needed to carry men and materials farther down and return to the base each night. A camp must be set up well down the river where we could live close to the working area.

From the Army Quartermaster Corps we got a large tent, rations, and cooking gear and set out on the evening of July 11 with hours of daylight still before us. Wengerd and I, with four enlisted men—Bettano, Cifizzari, Faretra, and Handschumacher—piled our gear into the M-T boat and sped down the river in search of a camp site. Some 12 miles downstream near Big Island, we fancied we might find an anchorage on the west side of the river, and started exploring for one.

We had scarcely begun our quest when the falling tide let us down on a cluster of boulders concealed from view by the muddy water of the river. The boat settled comfortably in a natural cradle between boulders, immovable as the further ebb of the tide revealed more and more boulders all around us in the deepening twilight. Wengerd and Cifizzari took their sleeping bags and rowed ashore in the skiff that we towed, to examine some likely-looking terrain for a camp site. They found a comfortable nook where they slept till morning, while the rest of us stayed on the M-T boat. The night was calm and mild and there was no hardship in our lot, except that the boat offered no comfortable place to lie down and sleep. About 1:00 A.M. when the tide began to rise we put out an anchor, and then I stuffed myself part way down a hatch in the stern and toward sunrise managed to doze a bit.

After sunrise, when we had floated off and were riding to the anchor, we retrieved the others and renewed our search for an anchorage. We were fortunate in finding one close to the south shore of Big Island, well out of the river current. Rowing ashore, we found a camp site much to our liking. It was a level place with a soft carpet of caribou moss close to a small lake of pure fresh water. A few scattered evergreen trees provided just enough firewood for cooking, and an enchanting group of shallow pools of fresh water among the rocks reflected the sky and added charm to the setting.

The day was fine and we were eager to get on with the survey, so stopping only to cook and eat a bit of breakfast, we deferred pitching our tents till we had put up all the survey signals we could on the island—a tripod, a secondary station, and a white square painted on the face of a cliff.

Before leaving the base I had raised the question of cot beds, but the officer who issued the tent said, "There's no better bed than caribou moss; it's better than any cot."

I had been taught in my boyhood to dig a trench around a tent to divert rain. I thought of it now, but the weather was fine, and there was so much to do playing scoutmaster and cooking meals for these denizens of the city streets, dressed as sailors, wholly unaccustomed to camp life, that I let it slide. For two days I enjoyed the role of coxswain of the M-T boat, running up and down the broad river, landing working parties and their tripod lumber, trying to teach the boys to row the skiff, and taking notes on the tide. One day, instructing a peculiarly inept pupil at the oars, I gave the order, "Hold your port oar; pull starboard." He let the oar handles rest in his lap and replied helplessly, "I haven't the coördination."

The conclusion of our first four-day stay in camp on Big Island and the price we paid for my neglect in not digging the trench is best told in an extract from a letter home in the following chapter.

15 Camp on Big Island

In Camp, on Big Island
July 15, 1943

"Sherm Wengerd and I are cooped up in our little trapper's tent-for-two, with an icy gale shaking it as a terrier shakes a rat and the rain beating on its walls like machine-gun fire. Last night it rained so hard that it found leaks at the corners, so that in spite of the supposedly watertight floor, Sherm's blanket under his sleeping bag turned into a swamp. With my pneumatic rubber mattress I fared better, and only the last 8 or 10 inches of my sleeping bag, where I normally park my feet, got wet from rain coming under the edge of the door. To stop the leak near my head I used my towel as a dam and it is now wringing wet. Most of our clothes and belongings are either wet around the edges or soaked through. You may observe that since I started writing it has dripped from the roof of the tent onto the paper.

"In the big Army tent the four enlisted men haven't stirred from their sleeping bags, though it's nearly noon. Sherm and I breakfasted in this tent on Army breakfast rations, cold, for lighting a fire, though physically possible, struck us both as 'one too many.'

"We are camped on a bleak island in the river, with lovely rocks, pools and mosses, charming in sunny weather. The Army M-T boat we are using is anchored off, and having only a 24-pound anchor, I tied a 50-pound rock to the anchor line last night, for the weather looked very threatening. I'm profoundly glad I did, for the wind is pretty strong. Our skiff is pulled up on the rocks, but we'd hate to have to make the 12 or 13 miles back to camp without the motorboat. (There's now a lake 24 x 12 inches x ½ inch deep just inside the door of the tent.)

"Tuesday and Wednesday were full and busy days putting up survey signals. We all worked hard and got a lot done. This morning S. and I agreed the weather was just too bad to go out on the river, so here we are holed in, and it's really very comfortable in my sleeping bag, in spite of the water all about. Two working days should finish our immediate task; so I'm hoping we shall get back to the base Sunday morning, and I wonder what news we shall hear from Sicily and elsewhere then.

"Just before this last heavy shower Sherm and I put on our clothes and went for a short walk and brought in some driftwood to eke out our rather meagre supply of firewood. It helped to get our blood circulating, but we got back to the tent just in time to duck a deluge that would have nearly drowned us.

"I wish you could have seen the moon reflected in the little pools in the rocks and the pink sunset over the river night before last. It was heavenly. But this: I'm content to have you read about it.

"7:15 P.M. The gale increased during the afternoon till I should not have been surprised to see it blow my little tent down, but still the boat's anchor held. It has moderated just a trifle now, but is still pretty fierce. We have just finished our third meal of cold Army rations. No fire. Bettano complained of pains in his legs from a damp sleeping bag. I gave him two of Jim Ayer's better-than-aspirin tablets and a capsule of Nembutal. The others seem to be O.K., but I bet they are homesick.

"July 16, 9:20 A.M. I am now writing in the comfortable house where I stayed with E. R. and party just two years ago (July 12-14 to be exact). Just after I stopped writing last night, things began to happen fast. It was 8:00 P.M. and I was planning in fifteen minutes to crawl out of my sleeping bag for the last time, put on my boots, and have a last look at the boat before turning in for the night. Sherm had already turned in. Suddenly there was a shout from Faretra that the boat was drifting up the river (the gale blowing that way). On hearing Faretra's warning shout, Sherm yelled, "All up, general quarters!" I pulled on my soggy barka boots as quick as I could and most of my "waterproof" clothes. I told Bettano (the only one who is half a sailor) and Faretra to come with me in pursuit of the boat, and Sherm and the other two to start packing up sleeping bags and clothing, for we decided on the spot that, as long as the boat had mobilized herself and got us out of our sleeping bags, we would head back to the base, get dried out, and await better weather.

"I ran down to the shore and was amazed at what I saw. Our skiff had been left pulled up on the rocky shore with her painter tied around a heavy rock above high-water line. But the gale had driven so much water into the river that it was up about three or four feet above H. W. line. The skiff was well off shore with her painter disappearing into the watery depths. There was nothing for it but to wade out to her, which took me in nearly to my hips. I tried to pull the painter loose, but I couldn't budge the rock it was tied to, so I chopped the painter off with an axe which had happily been left in the skiff. Then we hauled out said skiff to dump out the gallons of rain water. Meanwhile, thanks to the 50-pound rock on the anchor warp, the motorboat was dragging very slowly away. If it hadn't been for the extra high tide, I don't think she'd have dragged at all. The three of us rowed out to her without difficulty, for she was down wind only about 250 yards. The rock was so heavy to heave up that Bettano concluded the line had 'caught on the bottom,' but I told them to pull like hell and they finally got it up. Then I ran the boat back to within twenty yards of the shore where the shelter made the water smooth, and told Faretra to row the skiff in for the others. Before he could get his oars mobilized the wind had blown him out 20 to 30 yards, and then he couldn't row straight into the wind to save his neck, he just zigzagged and got nowhere. After about five minutes of total failure, he was exhausted, and I took the motor boat alongside the skiff and picked him up. Then I left Bettano, who knows just enough to run the motorboat, in charge and rowed the skiff in myself. We packed up all the sleeping bags and clothing, left the food and mess gear in the big tent, took down the upright of my little tent, weighted it down with rocks, loaded the skiff, and the four of us rowed out to the motorboat and started up river with the skiff in tow.

"As soon as we got out of the lee of our island we struck a tide rip much like that off Uncatena Pt. in a northerly blow and west running tide. We were running right before it, and once a wave caught us, the stern swung round rather sharply, and all hands, including Sherm, not knowing the action of waves in a tide rip, thought I had done it and suggested that I shouldn't turn her so suddenly, for the sharp swing yanked the skiff broadside into a wave, so that she scooped up a big load of water. Later I gave Sherm a trick at the wheel and a chance to learn what a following tide rip will do if you're not quick with the rudder, but at the moment our big problem was the skiff. I told them to pull her up on the stern, which took all five of them some time to do, using all their strength. Then her stern, heavy with water, hanging out over the motorboat's stern, became the tail that wagged the dog, and when a wave started to spin us, the skiff's stern swung out to one side and trailed in the water, scooping up more from the top of the waves. I ran the boat slowly till they managed to get the skiff transversely across the stern and balanced amidships; then they could bail her out and lash her down, and all went well.

"When we started we were planning to go right through to the base, but as the twilight deepened and the water continued rough, we put in at Chimo and the good Mr. Stewart took us in, gave us hot coffee and cake, and fixed up beds for us. In the warm Hudson's Bay Company blankets we slept so soundly that when Mr. S. called us for breakfast, I had to shake each man hard to waken him.

"Returning to the big tent yesterday, if an undertaker had gone in there he would have cried, "Tally ho!", rolled up his sleeves, and swung into action in jig time. I've never seen a bunch of men so like hibernating bears; they sure can sleep!

"Tomorrow we plan to assemble some more gear to make the camp more comfortable, in case of another storm, and on Sunday return to it to finish putting up the survey markers in that region—about one and a half days' work. Then others of the gang will go down to triangulate from those stations in preparation for the sounders. They will use the camp in their turn and we may have to set up another camp further down.

"The officer who issued us our Army rations, mess gear and tent, is a rather comic fellow. When he told us not to bother with cots, we took his advice. When we got back today, I said to him, 'If you tell any of the boys who slept in that tent that caribou moss is better than a cot, you'll have a fight on your hands, regardless of age, rank, color, creed, or sex.'

"He replied, 'Caribou moss is all right if it doesn't rain; the trouble is I looked in Lydia E. Pinkham's almanac, and I turned to the wrong page!' "

16 The *Morrissey* at Chimo

The comfort of the base, mail from home, bright sunshine to dry out all our gear, and a real Finnish steam bath made the interval between camping trips a welcome respite. Our return to Big Island was delayed because of the forecast of a bad storm. The man who called us "sissies" for heeding the warning got the last laugh when the storm never came. On July 19th with five enlisted men, cot beds, and a stove that could be used in the big tent, Wengerd and I went down the river again and put our camp in order, prepared for the worst.

Up went the "twisters" along the river bank as fast as we could land the working parties, retrieve them again, and move them on to the next favoring headland. The boulder trains that line the deep channel in midstream proved a formidable menace, the more so since the heavy rain had made the water turbid with mud, and rocks could not be seen at a depth of more than 3 or 4 feet.

Our second day I landed two working parties. One of these was put ashore at Stewart Point, a long promontory extending far out into the river. On the shore I met Tommy Gordon, an elderly Eskimo holding a position of eminence in the Chimo community and the only one who could speak English. The tide was now high, but falling fast. I asked Tommy if the sheltered cove north of Stewart Point afforded a good anchorage at low tide. His reply proclaimed foul ground where at low tide the motorboat would lie on her bilge for hours on the mud-flat strewn with boulders. Leaving the skiff on shore for the workers, I set out alone in the M-T boat to seek a way through the boulder train that would save them rowing out to the swift waters of the deep channel. The shallows extended for half a mile out until they terminated in a rampart of boulders at the edge of the deep water. Back and forth, back and forth I ran the boat, sounding with the lead till I found a passage with 8 feet of water. The M-T boat drew nearly 4 feet; could I follow this channel by ranges on the shore? Would there still be 4 feet in it when the men arrived in the skiff? Some of the higher rocks were beginning to show with the strong ebb current pouring over their summits. Soon I was joined by the others, who rowed out in the skiff and clambered aboard the motorboat. We started out, keeping the range points on the shore in line, but at the worst of the boulder train the current must have carried us off the range or else the tide had fallen faster than I expected, for we bumped on a rock; we were surrounded by them. Better make a dash for it, as the deep water was only a few feet ahead, and retreat would mean a wait of hours. I gave her nearly full throttle ahead and she jumped from rock to rock like a chamois of the Alps. Just at the edge of the deep channel where the current was setting us sideways with increasing speed, we fetched up on a boulder with a horrid crash, and heeled way over till a bag of cement and other things tumbled about in chaos. We slid over it, caromed off another rock, and shot out into deep water at last. The boat still ran, but there was an ominous vibration and her speed was reduced so much she could no longer buck the ebb tide at full strength.

For two days we continued setting up signals, planning our campaign to use the tides as best we could in our limping craft, and by noon on the 22nd we had completed our net of signals as far as Latitude 58° 22′, 13 miles down the river from Fort Chimo. Then we broke camp and on the flood tide returned to the base. On that day we got a good view of the boulder train below Stewart Point at dead low tide. It was an appalling sight. It looked as if half the stone walls of New England had been piled up by a steam shovel in one solid rampart like the Maginot Line. And that was what I had made the M-T boat climb over; no wonder both propeller and shaft were bent! Fortunately new ones were on hand at the base and our faithful boat was as good as new.

Whether from the stress of camp life or from other causes, four of our enlisted men developed complaints for which they had to be flown to the Chelsea Naval Hospital. I wrote the Bureau of Personnel. I was inclined to ask for a battalion of physical giants, all guaranteed proof against piles, psychoses, neuroses, headache, heartache, backache, homesickness, malaise, sciatica, lumbago, rheumatism, germs, toxins, muscle strain, and mosquito bite, but I figured we'd just get "the run of the mine," so I simply asked for replacement of the four hospitalized ratings.

During the days after our return from camp, Turner and I went up in one of the two PBY planes that the Navy had furnished for this mission and circled low over the river bank photographing our shore signals at close range, that we might place them accurately in relation to the shoreline on the chart we were to make. In these flights it was interesting to learn from the startled reaction of the Navy pilots that a slight side-slip, one of the safest of all maneuvers in a light plane, is dreaded as fraught with danger in these clumsy monsters of the air.

Before the end of July, contact was made with the first ships of the season endeavoring to reach the Koksoak River by sea. The PBY planes, flying far out over the pack ice in Ungava Bay, scouted in search of them. Four ships were on their way, the famous *Bear,* again flagship of the Canadian Arctic Task Unit, the buoy-setter *Laurel,* the *Belle Isle,* and, last and least in size, but foremost in importance to us of the Hydrographic survey, the schooner *Morrissey* with the indomitable Captain Bob Bartlett, coming to fetch us. On July 27 we got our first radio from the *Morrissey* somewhere out in the vast ice floes of Ungava Bay. Puett flew out in his PBY and searched for her in vain—an infra-needle in a super-haystack.

Next day more messages indicated that the ships were getting nearer. The tension of eager expectancy increased. On the 29th I went down the river in a motorboat with Lieutenant Maier, Naval Port Liaison Officer, and Edmunds, the veteran chief pilot of the Koksoak River. We ran right out into Ungava Bay and there found the *Laurel* and the *Belle Isle* at anchor near the river mouth. We went aboard the *Laurel,* a fine new ship, and were cordially received with appropriate refreshments by her skipper, Commander Lok, in his cabin. Neither the *Bear* nor the *Morrissey* was yet in sight. On the late afternoon flood tide the *Belle Isle* led the first procession of the season up the river to Fort Chimo. Maier had already learned the channel and was able to pilot a ship safely up the river, a task hitherto attempted only by Edmunds and his Eskimo colleagues.

Next day after lunch Maier and I went aboard Lieutenant Tomlin's PBY and flew down over the river in search of the *Bear* and the *Morrissey.* To our surprise we saw the *Morrissey* just entering the river mouth, trying to buck the strong ebb tide with her engine. That no pilot had come aboard and that no one save the ill-fated tugboat captain had ever attempted to take so large a vessel up that dangerous stream since the first explorers learned of its terrors was no deterrent to Captain Bob; it took more than the Koksoak River to stop him. I was eager to board the *Morrissey* and help him, for already I knew the location of the dangerous boulder trains, but Maier's first objective was the *Bear* and we must fly on in search of her. We found her at anchor some miles off the river mouth with the trawler *Aklak* near by. Thick weather balked our efforts to take the latter up to Chimo till the evening of the 31st when the flood tide brought us quickly to anchorage off Chimo. There we found the *Morrissey* and her famous skipper. Captain Bob, a total stranger to this river, had done what probably no other captain would have dared—trusted his ability to guess the channel—and had pushed on up the river as soon as the ebb tide slackened. He guessed right most of the way, but just south of our camp on Big Island he guessed wrong, bumped on two or three boulders, but was carried over them by the strong flood tide, his stout keel protecting the schooner from damage.

Busy days followed as we prepared to head northward for Frobisher Bay. More ships came up the river bringing heavy equipment that needed to be sorted and stowed.

On August 6, the eve of our departure for Frobisher Bay, a U.S.O. troupe arrived by air to give an evening entertainment to the personnel of Crystal One. Paul Draper, the noted tap dancer, two young ladies who danced and sang, and a master of ceremonies made up the band of entertainers. A cargo plane set out from Goose Bay carrying a piano, but was forced back by bad weather. Learning of this by radio, Draper rose nobly to the emergency and performed admirably without music.

The next afternoon Wengerd, Mohl, Turner, Price, Blasey, and I went aboard the *Morrissey* to set out for our survey of Frobisher Bay. At 3:00 P.M. we started on our long voyage. I had surveyed so much of the river by now and had watched Maier at his pilotage enough to feel easy about piloting the *Morrissey* down to the sea. To my great relief Captain Bob, who had been rather touchy about suggestions the year before, now accepted my role of pilot with the docile acquiescence of an obedient spaniel. Long before dark we had left the Koksoak River far behind us and were gliding smoothly over the broad expanse of Ungava Bay.

17 Back to Frobisher Bay

We encountered pack ice our first night out and all next day made slow time zigzagging through the lanes of open water and ramming the ice pans as we had done the year before in Frobisher Bay. It was an old story to the *Morrissey's* crew, but a startling novelty to our hydrographers. In the twilight the Captain decided to tie up to a large ice pan for the hours of near darkness. This was a maneuver I had not seen before. Running slowly to the edge of a pan of some 2 or 3 acres in area, he stopped the ship as the bow touched the ice. Two of the crew dropped from the bobstay to the ice, carrying a small anchor and a pickax. Running a few yards up the icy slope of the pan, they cut a deep hole with the pick, drove a fluke of the anchor into the hole, then climbed back aboard. With the warp pulled taut, the ship lay snugly secured in the lee of the pan until the dawn brightened enough for further progress through the floe.

Familiar landmarks that brought back memories of the previous summer passed by as we headed northward. By noon of August 11 we had arrived at Bartlett Narrows where in September we had done some of our most intensive work in locating and plotting the reefs that guard the channel. We were duly met off Peak Island by an M-T boat with a new pilot aboard. I hailed him and asked if he would like us to pilot him through the Narrows. He took it in good part and on we went, sounding with the fathometer, zigzagging between Scylla and Charybdis as of yore, passing Pink Lady, and soon after six o'clock dropped anchor in our familiar berth in Koojesse Inlet. The base had grown to look like a populous village and from it Major Davies, the chief Engineer Corps officer, came aboard to welcome us.

As in the previous year, the bay had been open since the first of August, and now the weather was already beginning to turn cold. We had missed ten days of the best weather for starting our survey, days that we could have used to good advantage had it been feasible to mobilize on the spot as soon as the ice went out. Now our survey crew went to work with all speed, dividing into squads dispatched by whaleboat to seek the high spots on headlands and islands for primary stations in our triangulation network. By the end of the first day three tripods had already been erected in the vicinity of Koojesse Inlet. Canadian surveyors had already made an astronomical determination of the precise geographical position of a marked point on the hillside near the base, and had given us their data. This point we named "Astro," and it became the key to our whole survey.

As on the Koksoak River, so in Frobisher Bay the plan was the standard procedure for a hydrographic survey—to erect tripods that would form a triangulation net or skeleton for the chart, then locate secondary markers along the shore, triangulated from the primary tripods, and then to take lines of soundings placed by reference to the shore markers. The width of the bay and the numerous islands therein necessarily made the net more complex and extensive than the simple double row of tripods down the sides of the river. But with the *Morrissey's* two whaleboats at our disposal we made rapid progress putting up tripods around the head of the bay. The great island barrier dividing the upper bay from the much larger lower bay presented fascinating problems. We had to climb to the heights of many of the islands in search of key points from which we could see the requisite number of other tripods. Sam Bartlett and I traversed the island where Hall was pursued by wolves, but no such terror disturbed our peaceful walk. Our quest led us to a lovely snug harbor on the north shore of Allen Island, thence up a trail of sorts passing close to the shore of a large lake high above the sea and on up a ridge, where the way was studded with Arctic poppies of delicate pale yellow tint, to a commanding summit. "Here," I said, "is where I'll make my Frobisher Bay home." But I doubt if I ever shall.

The number of islands taxed our ability to find names for all of them. When possible we adhered to the names given by Hall, but in many places the islands differed so markedly from what his map showed that identification was impossible. One day Mohl, Blasey, and I landed on Long Island in Koojesse Inlet to triangulate other stations from the tripod on its summit. The whaleboat was to have come for us before lunch-time, but while we were at work the crew went in her to the mouth of a brook to fetch a supply of fresh water and got stuck there on the falling tide. Thus marooned on the island, we had time to spare for exploration and found an ancient Innuit structure of stone, presumably a fox trap. Fortunately the radio man on the *Morrissey* signaled by semaphore to another boat that brought us a dory in which we could return for a belated lunch.

The next day, August 19, we were scheduled to set out in the *Morrissey* on a threefold errand. First there was the transportation of Paluchi and all his family, dogs, worldly goods, and household gods. Paluchi was the versatile and obliging Eskimo who was Crowell's "factotum" throughout his year at the island base down the bay. Crowell found him most helpful, and on leaving Frobisher Bay bequeathed him to his successors. But, alas, when Crowell left, no one remained who understood the Innuit, and the Army personnel expunged him from the Government payroll. Now they wanted to ship him and his family back to his old hunting ground on Crowell Island.

Major Davies, having no suitable boat available, asked us if we could take the party down, eight people, about ten dogs, and all their goods, including a winter's supply of flour. Then the skipper of the steamship *Eagle,* scheduled to leave Koojesse Inlet for the south, asked if we could pilot him out through Bartlett Narrows.

Waterfront at Crystal One, Koksoak River, July 1943

Camp on Big Island, Koksoak River, July 1943

Eskimo boys at Chimo, 1943 (note one with Anglo-Saxon features, unaware of racial difference)

Erecting a tripod, Frobisher Bay, 1943

Filling water casks, Frobisher Bay, 1943

Lunch hour in the MORRISSEY'S *whaleboat, Frobisher Bay, 1943*

Major Hassell, Captain Bartlett, Lieutenant Gregg, Commander Forbes, Frobisher Bay, 1943

Sounding boat NANUK *in Peak Island harbor, Frobisher Bay, September 1943*

Wengerd examining Eskimo fox trap on Long Island, Frobisher Bay, 1943

Kodlunarn Island, showing "ship's trench," September 1943

MORRISSEY *crossing the bar in Cincinnati Press Channel, October 1943*

Homeward bound on the Morrissey, *October 1943, Turner and Wengerd on the rail at the right*

Grinnell Glacier, Meta Incognita

These two errands could be combined with survey operations in the islands down the bay. So it was all arranged that Paluchi and his tribe should come aboard at 7:30 A.M. and we should lead the *Eagle* out through Bartlett Narrows, then cross the bay to Crowell Island and land the Eskimos there.

Bright and early, a barge came alongside and dumped about a ton of Innuit baggage on our deck. Seven-thirty came, then 8:00, and our binoculars revealed no trace of approaching Eskimos. So Captain Bob said he would start out of the harbor—that would make them hurry. Slowly we went out and when 8:30 came and nothing had stirred on the beach we decided that we shouldn't keep the *Eagle* waiting any longer. So we put on speed and went on our way with the baggage but not its owners. We stopped to set Turner and a crew ashore on Pink Lady to erect a tripod and picked them up on our return from speeding the *Eagle* on her way. Thence we crossed to Crowell Harbor where to our amazement we found an M-T boat by the shore and a tent on the beach. This little 28-foot boat had brought Paluchi and his kith and kin, dogs and all, 26 miles down the bay from the base. They had been ready to sail with us on schedule, but Major Davies had been unable to procure a boat to bring them aboard the *Morrissey,* and only after we had gone had he succeeded in mobilizing the boat. Not only had she brought eight Eskimos and as many dogs across the rough waters of the bay, but another ton or so of supplies. How they all got into her was more than we could see.

As soon as we anchored, the M-T boat came alongside and transferred this additional cargo to our deck, for they had no means of landing it on the beach at that tide, and they had to get back to their base before dark. The problem now was to get the double load of Eskimo dunnage ashore. Captain Bob felt that he had already done enough for Paluchi, and his whaleboats were needed for our work. An old dory was fetched from the shore and filled with a first load. Paluchi paddled out in his kayak and tried to tow the dory to the beach. The wind was so strong against him that he could barely make any progress. It is hard to discourage an Eskimo, and for some minutes we watched him, patient and uncomplaining, as he paddled valiantly toward the distant shore. At last Captain Bob's compassion prevailed and he ordered the whaleboat to take Paluchi ashore, towing the loaded dory.

The trouble Major Davies had in securing a boat to meet his schedule was an example of the difficulty of coördination between the Engineer Corps and the military command at such a base. A more costly instance occurred later when the Signal Corps personnel needed a heavy piece of radio apparatus brought from Crowell Island to the new base at Koojesse Inlet. Major Davies had the necessary barge available for the task, but although the officers lived in adjacent houses, it was impossible for the Signal Corps personnel to get a message to Davies through official channels. Long delays and many headaches ensued before the job was done.

18 A Trip to Greenland

On August 24 we returned to Koojesse Inlet after a six-day tour of the barrier archipelago all the way to Daniel Island, busy days of exploring and putting up tripods on many islands. At the base we found that a freighter had just arrived from Boston with our 32-foot sounding boat aboard. About the same time three big airplanes arrived, all equipped with surveying cameras sent by the Hydrographic Office to photograph Frobisher Bay with both vertical and high oblique shots and thus provide the makings of a map to be built on the basis of our triangulation. A perfect flying day, clear and calm, afforded Lieutenant Gregg a chance to take Captain Bartlett, Mohl, Turner, and me up in the Beechcraft for an informal survey of the upper bay. For more than two hours we cruised around at our pleasure, and most of that time Gregg let me fly the plane—a smooth and responsive craft in spite of its big size. The Navy had played Santa Claus and provided a vast quantity of Kodachrome film with which we could take scenic views and experiment at recording the position of shoals by the color contrast of the brown kelp with the dark blue of the surrounding deep water. Captain Bob, sitting in the nose of the plane with only plexiglass between him and the wide panorama before him, drank it all in with the enthusiasm of a boy.

My functions as "officer-in-charge" of the survey were somewhat nebulous, based apparently on seniority and on previous acquaintance with these northern waters and with Captain Bob. Having in each area a trained hydrographic surveyor designated "chief engineer," who knew the technique far better than I did (Stirton at Chimo and Mohl in Frobisher Bay), I had been specifically instructed to leave the operation of the survey to them. I just lent a hand when and where I could.

One day a real task was put up to me; Wengerd and Turner came to me with the complaint that the morale of the survey crew needed jacking up. One man complained that as a diesel engineer he found the gasoline engine in the sounding boat, *Nanuk,* unworthy of his best efforts; he was wasting his talents and not getting on with his career. Another objected to being taught the details of his job by another enlisted man with no more rank than he had. Frobisher Bay was far from the fighting front and our survey was too trivial to bother with, and so on.

We assembled the survey crew in the *Nanuk's* cabin and I let loose on them such oratory as I could command. Frobisher Bay was indeed far from the fighting, and I was painfully aware that for all the eagerness the men at the very top had displayed in 1941 to get these airports built quickly and used to dispatch flocks of fighter planes to the front, and for all the millions spent building them, now in 1943, with the airports built and ready, not one plane had been ferried to Britain by this northernmost route. But Admiral King had sent us to survey the bay and valuable ships were coming and going; it was easy to make a case for our obligation to carry on. To deal with the snob I recalled the case of naval officers at Liverpool in 1918, listening with due humility while an enlisted man taught them the workings of an antisubmarine listening tube. As for the man who despised the gasoline engine, I predicted that it might yet pose a problem that would challenge his best engineering talents, and later it did. There were no shouts of applause, no exuberant battle cries when I concluded my remarks, but anyhow the survey went on.

On our next cruise among the islands in the *Morrissey* as we reconnoitered the waters west of Bartlett Narrows, dense fog rolled in from the lower bay, choking the Narrows and obscuring both its shores. Three freighters came steaming down the bay from Koojesse Inlet, led by the station boat, *Arluk,* bound out to sea via Bartlett Narrows. Into the fog they plunged and disappeared. The sight gave me the biggest thrill yet, for it meant that our reconnaissance survey of the previous year had provided data reliable enough to enable the skipper of the *Arluk* to zigzag through the passage between Scylla and Charybdis, virtually with his eyes shut, and take this convoy of big ships safely on their way to the open sea.

Returning again to Koojesse Inlet on September 4 we learned that the buoy-setter *Laurel* had arrived during our absence and her crew had put up lighted beacons on Monument and Long Islands to guide incoming ships to the anchorage. The *Laurel* had brought a load of buoys to mark the channels in the bay, but her skipper, Commander Lok, had just received orders to proceed at once to Iceland, stopping only to refuel at Gröndal, in southern Greenland. He had therefore put all the buoys ashore at the base. These buoys were heavy spars and only the *Laurel* had the gear needed for mooring them.

There were three reefs that we greatly wished to have buoyed —Scylla, Charybdis, and another that we called "Deception," drying about half tide and dangerously concealed at high water. All three lay close to the course from Bartlett Narrows to Koojesse Inlet and offered real hazards to ships. Of course the buoys we had put out last year had been swept away by the winter's ice.

I went aboard the *Laurel* to see the skipper. He was most cordial and though he must leave for Iceland next day, he offered to buoy these reefs on his way to Greenland if the buoys could be put aboard in time and if I would go with him to show him where to set them. Here were two problems: how to get the buoys lightered from the beach to the *Laurel* and how to go down to the Narrows without being carried on to Greenland. I might get the *Arluk* to come down and take me off the *Laurel* before she sped on her way; but suppose we missed connections? Well, what if I did go to Greenland? I had always wanted to, and that was the headquarters of Admiral Smith whose advice I would do well to seek about the future conduct of our survey. Perhaps I needn't bother the *Arluk* to come after me.

I talked with my colleagues on the *Morrissey,* and all agreed that the survey routine was now well organized and my absence for a week would make no difference; a conference with Admiral Smith would be timely. Airplanes were flying almost daily from Narsarssuak to Goose Bay, and I could be back in Frobisher Bay almost before they noticed I had gone.

Next was the tougher problem of getting the heavy buoys off the beach and aboard the *Laurel*. With all speed I found Balkin, a real man of action. Most men would have said it couldn't be done. The tide was rising fast, and soon would float the grounded lighter. Balkin mobilized a truck in double-quick time and got the buoys loaded and down to the lighter just before the rising tide chased the truck back to dry land. The lighter ferried them to the *Laurel* just in time before she was scheduled to sail.

Down the bay we went, stopped and felt our way cautiously to the edge of Deception Reef, sounded the depth of water, measured off enough chain to moor the buoy, shackled it, and then the sturdy crane swung buoy and concrete mooring over the side and let them go. On we went to a quiet anchorage by Pink Lady Island where we stopped for the night. Next morning we set out the buoys by Charybdis and Scylla, and went out through Bartlett Narrows and down the bay where the afternoon sun shone full on the matchless splendor of the Grinnell Glacier, and out on the open sea, bound for Greenland, a two-day journey.

The *Laurel* was a fine new ship with the latest gear for all her needs, not to mention comfortable quarters and a good library. Her voyages in these same waters two years later have been described in an entertaining book entitled *Red Flannels and Green Ice,* by her Executive Officer, Arthur Pocock. When I was on her the crew vied with each other in spectacular beard-growing contests. From the callow features of a youth in his teens, a ferocious black beard sprouted bravely to belie his tender years. There were regal beards, Biblical beards, the Mephistopheles beard, the Henry VIII model, and many others.

On the deep blue sea fulmars were all about, soaring with superlative grace, their wing-tips tracing the contours of the waves, missing the surface by a hairsbreadth—maneuvers that would terrify a seasoned aviator.

About eight o'clock in the morning the bridge reported that the sonic detector had picked up two objects ahead—maybe submarines. Dashing up on the bridge, the skipper sounded "General Quarters." All hands were alert and scanning the horizon in the direction where the dreaded objects lurked. Binoculars were trained and presently two whales came to the surface a mile or two away and spouted. That was all.

On September 8 a tall, sharp peak appeared piercing the top of a fog bank, our first glimpse of Greenland; then the morning sun banished the fog and the ship slid through a narrow strait where the rocky shores seemed almost to graze the bilges on both sides. Before noon we entered a lovely fiord and docked at Gröndal. There we found MacMillan's famous schooner *Bowdoin* secured to the wharf. With a record of many hundreds of miles of Greenland waters behind her, she was then serving the Navy by charting these fiords. Ten minutes after we had docked, our radio told us of Italy's surrender.

Next day with full tanks, the *Laurel* threaded tortuous passages by frowning peaks and shining icebergs towering above us, around Cape Desolation, and brought us in the twilight to the head of Narsassuak Fiord where the long runway of Bluie West One slopes up from the shore toward the foot of a glacier. Here I motored past an enormous and populous base to the Navy quarters, where I learned that Admiral Smith was far away. I couldn't confer with him, but anyway, I had seen Greenland.

For three days I awaited flying weather that would get me back to Goose Bay. Marooned in this austere valley brightened only by the rich red of the birch leaves, I read *Lost Horizon* and compared my lot with that of the Englishman marooned at Shangri-La. This base was near enough the fighting lines to call for a degree of readiness beyond anything I had seen at the Crystal bases. All the Army men came to the movies in the evening with rifles slung over their shoulders. During my stay Mrs. Hooper and two other aviatrices came in as copilots of planes bound for Europe. When they appeared in the audience at the movies, the men stood up on their chairs, cheering and shouting at the top of their lungs; they had seen nothing like this for many months.

At last the weather improved. A beautiful flight down the fiord, across the sea and over the familiar shore of Labrador, a night at Goose Bay, and two flights via Crystal One brought me back to Koojesse Inlet and aboard the *Morrissey* by the evening of September 14.

19 Homeward Bound

The survey party had been working hard and now had complete soundings of Koojesse Inlet and approaches. Before leaving to sound Bartlett Narrows, Turner and I were taken up for an extensive flight over the entire length of Frobisher Bay. Lieutenant Alley took me in the PBY, while Turner went with De Garmo in the Ventura. We flew as far as Lok Land, taking many pictures. It was on this flight that I saw the "ship's trench" on Kodlunarn Island, which Hall reports was used by Frobisher's men to build and launch their boat in 1578. Viewed from 10,000 feet aloft, it looked just like the drawing of it in Hall's book of 1865. I took three photographs of it, one of which is shown in this book.

As we returned up the bay toward the base, clouds came down on the hills with quite unexpected speed. Descending rapidly we kept below them and landed with a comfortable margin before they closed in; but the Ventura was still up and not in sight. The radio room was apprized of De Garmo's concern lest the clouds come all the way down and cut off his approach to the field. For a moment he was on the verge of turning back and flying south 300 miles to Crystal One—a dreadful prospect, for Turner, leader of our sounding crew, was in the plane and his absence for even one day would be a severe setback to our survey; and then who could tell when the weather would permit his return? Officers ran to the roof and like Wagner's Valkyries on their rocky lookout, anxiously scanned the cold autumn sky, concerned more with the safety of plane and crew than with the progress of our survey. The suspense increased as the minutes passed, but at last, to our great relief, the Ventura broke through and came in for a safe landing.

The task of a hydrographic survey includes observation of the tides—the range of their rise and fall—and tidal currents. By the shore of Koojesse Inlet we set up an automatic recording tide gauge; wound up like a clock once a week, this wrote on a slowly revolving drum a record of the rise and fall of the tides. At Crowell Island and at Bartlett Narrows we used the simpler method of setting up a staff with feet marked on it. This meant detailing one of the crew to observe the water level with binoculars at frequent intervals and keep a written record.

The most remarkable observations were a tidal range of 38 feet at Koojesse Inlet on a spring tide and on the next neap tide a range declining to 11 feet. So large a difference in tidal range is unusual. Another surprising effect was a reversed direction of flow in one of the passages between islands in the barrier archipelago dividing the upper from the lower bay. These islands are mostly long and narrow, lying in parallel lines that run northwest and southeast in alignment with the long axis of the bay. Through these passages the current runs fast. One island whose outline resembles that of a tadpole with its tail curled up, defied identification with any of those shown on Hall's map. We named it Brigus Island after Captain Bob's home port, and made several visits to it for the erection of a tripod and triangulation from it when visibility served. Through the narrow passage between it and Allen Island, its next neighbor to the south, the tide ran with great force in the opposite direction to that in all the other adjacent passages. It seemed to be an eddy current. A glance at the map will show that the bay between Brook, Allen, and Brigus Islands could act as a pocket into which the tide flows simultaneously through the several openings leading in from various quarters. The apparent eddy current is probably to be explained in this way.

While the *Nanuk* was tied up at the base for minor repairs, the *Morrissey* preceded us to Bartlett Narrows, the next region due for detailed sounding. Captain Bob was always glad to get away from the hurly-burly of Crystal Two. Turner and I stayed ashore at the base till the *Nanuk* was ready, then on September 18 set out in her to follow and find the *Morrissey*. The morning was fair with a calm sea, and we set out expecting a smooth voyage, but soon clouds gathered and a strong southeast wind kicked up a choppy head sea and drove wet snow against the windshield. The weather was far more rugged than the boat, but still the engine ran and we were glad of the shelter from flying spray and snow. Fog closed in and we hugged the shore east of Sybil Island, glad that the shoal draft of the *Nanuk* enabled her to go in safety wherever no breakers appeared to betray rocks close to the surface. Soon the dark cliffs of Frobisher's Farthest loomed up ahead through the fog; we skirted its west shore, bucked a steep tide rip, picked up Alligator Island, found smoother water in Bartlett Narrows, and at last after a chilly voyage of four and a half hours slipped through between the reefs into the shelter of Peak Island Harbor, where we found the *Morrissey* in her snug berth. Captain Bob had little confidence in the *Nanuk* and seemed rather shocked that we had come down the bay in such a craft on such a day. Anyway, there we were, and glad to be there.

The next nine days we concentrated on sounding Bartlett Narrows and approaches with a thoroughness worthy of its importance. There was plenty for all of us to do. Turner with his sounding crew in the *Nanuk,* now well drilled in the standard procedure of the Hydrographic Office, acquired a proficiency that stood him in good stead when later he was assigned to a far more hazardous sounding job with the amphibious forces in the Pacific, under fire of enemy far more formidable than the Baffin Island mosquito or the chill of early winter. In retrospect I am convinced that in prying Turner loose from a trivial job in the Hydrographic Office and giving him this experience I made about my best contribution to the war effort!

To strengthen our local triangulation I landed one day on Mitchell Island to take a round of angles with a transit from a cairn we had built in 1942. One of our Navy boys went with me to take notes. Although he had lived all his life in a great seaport, this boy, so he told me, had never been on a boat till he came to Frobisher Bay, having spent his time mostly at "dances and that sort of thing." Mitchell Island was an intriguing jumble of bastions and ramparts of massive rock interlaced with gullies carpeted with mosses and arctic wild flowers. As we returned to the boat I remarked, "This is a fascinating island, isn't it?"

The boy looked at me in blank amazement.

"Don't you see anything in it?" I asked.

"No."

"What sort of island would you like, one studded with palm trees?"

"One with lots of women on it," was his reply.

Winding up our work at Bartlett Narrows in the closing days of September we all migrated across the bay to Crowell Harbor. There Turner lived ashore with his sounding crew in Crowell's abandoned camp, the original Crystal Two; and glad we were that the houses still stood there amply equipped for such a party. Here was our last main task—to sound Crowell Harbor and its approaches. These included the way out into the upper bay to the north, and Cincinnati Press Channel to the south.

On September 30 the *Nanuk's* rudder skeg broke where it joined the keel, and dangled from the rudder. Running at reduced speed Turner managed to do a day's sounding in her, but Captain Bob declared the boat a menace and said we had better get her back to the base at once. She was built for summer week-ends on Long Island Sound rather than for the rigors of Frobisher Bay. The sounding gear was transferred to the M-T boat, borrowed from the base, and on October 1 the *Morrissey* escorted the *Nanuk* back to Koojesse Inlet, where she was hauled out for the winter.

That evening I saw the most beautiful aurora I have ever witnessed. The first remarkable feature was its appearance not long after sunset when the twilight was still fairly light and full of color; into this sunset sky came the display of pale green edged with pink, flickering as only an aurora can. As we went out from the beach to the *Morrissey* in the deepening twilight, the aurora became more spectacular. I've often seen flickering and rapid shifting in these things, but never like this. The streamers of colored light whipped and danced and flashed across the sky with incredible speed; their motion compared with the ordinary motion of the aurora was like low clouds tearing past in a full gale compared with the lazy drift of high clouds in a gentle breeze. They looked like ethereal banners of the softest silk, shading from pale green to rose color, waving wildly in a hurricane. Calm as the evening was, it looked as if some tremendous blast were sweeping these intangible banners overhead.

Next day we returned in the *Morrissey* to Crowell Island, where Turner in our absence had sounded all of the harbor and most of the northern approaches.

Winter had now come to Frobisher Bay; snow lay on all the islands, and when the north winds blew it was a bitterly cold business to take our rounds of angles on the distant tripods. While Turner worked on the harbor and northern approaches the rest of us turned our attention to Cincinnati Press Channel. We had soundings of the bar across the channel recorded in July 1942, but though our fathometer had shown the rest of the channel to be deep, none of our other soundings had been properly located.

On October 3 we had a banner day. The *Morrissey* moved down to Eggleston Bay. Wengerd, Mohl, and I with several of the men set up marker after marker along the shores of the channel. The day was clear and so calm that in the bright sunshine we were actually warm as we worked, warm enough to shed our coats. By nightfall all was ready for the big day of sounding on the morrow. Turner still lived at Crowell's camp with the sounding boat anchored off, while the rest of us turned in on the *Morrissey*.

Toward morning I stirred in my sleep, dimly aware that the ship was not lying to her anchor as peacefully as in the calm of evening. But such was my confidence in Captain Bob and his schooner that I slept again. At seven I came on deck to find the sturdy vessel lurching and straining at her anchor chains. A hurricane was roaring down out of the mountains, tearing the tops off the waves and driving great sheets of spray horizontally before it. Captain Bob paced nervously on the quarter-deck watching the fury of the storm.

"How do you think the sounding boat is taking it in Crowell Harbor?" I asked.

"I'm thinking about the *Morrissey!*" he thundered.

In the calm evening he had put out both anchors with 105 fathoms of scope on each, and when the hurricane struck soon after midnight he had run the engine for a time to be sure the schooner didn't drag. When he told me this, I asked how he knew it was coming.

"It's time for bad weather in this country," he replied. "We've been having good weather too long. And then I could tell by the way the birds were talking."

If any man knew the signs and portents of the north, it was Bob Bartlett. The sea gulls had a message for him that he alone could understand.

"It's time to get out of here; things can happen fast now," he said; and what he said went. The wind had blown down half of the shore markers we had erected the day before, although some were stayed with eight or nine stout wires, instead of the usual four. All things considered, we quickly reconciled ourselves to the Captain's dictum, and that is why the chart shows Cincinnati Press Channel with only a cluster of soundings on the bar and long blank spaces on both sides of it.

As soon as the wind moderated enough we got under way. Eggleston Bay, which we had previously found calm when fresh winds were blowing in adjacent waters, now seemed to Captain Bob a danger spot when storms were brewing in the mountains. We moved back to our first anchorage by the north end of Pugh

Island whence we could converse with Turner on Crowell Island two miles away by walkie-talkie radio. We told him to break camp on the morrow and bring all hands aboard for our farewell journey to the base.

Next day the storm had abated, but a stiff west wind was blowing. We picked up Turner and crew in Crowell Harbor, and by 6:20 P.M. were in Koojesse Inlet, where Mohl and all the Naval enlisted men went ashore at the base to make the journey home by airplane. Wengerd, Turner, and I stayed with the *Morrissey* for the voyage to Brigus by sea. A day sufficed for our business ashore and our farewell to the base.

At 5:30 A.M. on October 7 the *Morrissey* was under way, bound for home. We digressed only to enter "The Lion's Den" and make a brief sounding run on which we located at last the submerged reef we had seen from the air, but hitherto failed to find. With this last morsel of information for the chart we went on down the bay, watching the towering crags of Meta Incognita fade into the wintry storm clouds that hid their summits. Strong winds from the north and west favored our journey home and with trysail, foresail, and jumbo set, the old schooner logged 9 knots—the best I had ever seen her do.

At noon on October 14 we docked at Brigus and bade farewell to the *Morrissey*. Catching an Army bus at Holyrood, Wengerd, Turner, and I were in Argentia by 10 P.M. and next day we flew nonstop to Boston in a Navy transport plane.

Looking back on my two summers with Captain Bob I was struck by the great difference between them. In 1942 he found the assignment harassing at times. Accustomed to be the undisputed autocrat aboard his ship, it seemed to irk him to be tied in any way to military demands. In 1943 he had mellowed to a remarkable degree. Signs of irritation were rare and humor prevailed in his mood. He told me once that his ancestors were Spaniards who came ashore from the Armada on the coast of England and settled there. Latin blood doubtless contributed to his quickness of perception and to his sensitive appreciation of human values. My friendship with him ripened to a rare possession that I cherished till his death in 1946.

20 Conclusion

On our return from the north to Washington, in October, we delivered the survey material to the Hydrographic Office. All the triangulation notes and many dozens of aerial photographs, vertical and oblique, of Frobisher Bay were turned over to the Division of Chart Construction, there to be put through the standard procedures for making such charts as were deemed feasible and expedient.

The first matter to engage my attention was the development project in oblique photogrammetry which had been farmed out to Merrill Flood and Associates by the National Defense Research Committee. The project had just reached the active stage when we went north in June, and now, at the end of October, there were reports on many conferences to be studied. Officers returning from the Pacific reported that in some important places the available charts were very bad; better charts were urgently needed. Oblique photographs were said to be much easier to obtain from the air than verticals, but thus far only verticals had been used for mapping, so they said, since means for mapping from obliques were not available. Here, then, was a tremendous challenge—to make available at the earliest possible moment some practical means for converting oblique photographs into maps in the field of operations, as well as in the mapping laboratories at home.

The committee and Dr. Flood's group at Princeton had concluded that Miller's proposed single eyepiece plotter would require too much time and skilled labor to be feasible under the stress of war; simpler and speedier means must be found. The most promising answer seemed to be automatic photorectification, either by pinhole or by a specially designed lens. The essence of the idea was to rephotograph the oblique picture by projecting its image upon a map plane through a pinhole or lens adjusted to the same angle of obliquity as that of the original photograph. Miller was again consulted and came forward with the most promising answer to the riddle yet offered—a two-stage rectifier in which conversion from oblique perspective to the true map could be done in easy stages that would make less severe demands upon the lens. Intensive work was begun on developing such an instrument, and while that was in progress a pin-hole projector was made in Philadelphia. The precision sought in this instrument was a source of much delay in its completion. Could a simpler, cruder rectifier be made in a hurry and still be good enough for approximate mapping in the field? This question exercised me, and I finally managed to get a simple and portable pinhole photorectifier made in the Pattern Shop of the Gun Factory in the Washington Navy Yard. Experimenting with this in the darkroom, rectifying obliques from Frobisher Bay and comparing the results with vertical photographs of the same area, furnished an intriguing occupation during intervals between conferences in which the more formal work of the committee was carried on.

The importance of the northernmost air route via the Crystals and Sondre Stromfiord had been steadily waning since the first burst of energy, when the airports were planned and built. Reconnaissance flights over the Greenland icecap had resulted in losses of planes and men, and this had bred a great fear of that route, especially if flown by the young and inexperienced fighter pilots who would have to take the planes across. The northernmost route looked formidable. Some P-38 fighter planes were equipped with extra fuel tanks that enabled them to make the longer hops from Goose Bay to Narsarssuak and thence to Iceland. Our air and surface patrols were fast gaining the upper hand over the enemy submarines until it was feasible to ship the short-range fighter planes on fast freighters and to do so with less risk than was involved in flying them across Davis Strait and Greenland. And so it came about that the northernmost air route, deemed so vital by the High Command in 1941, was never used; and the airports along the way, built at such prodigious cost, found their main utility in the conduct of the war was that of weather observation—a valuable contribution to be sure, for on the weather reports from the far north depended much of the safety of air operations farther south.

This outcome led to a sense of futility and dejection. But I found consolation of sorts when I saw the gunners standing by their antiaircraft guns on the banks of the Potomac, waiting patiently for the German bombing planes that were to have bombed Washington but never came. Comparing my lot with that of these gun crews, I realized that if my job had been almost as barren of military achievement as theirs, at least it had been a whole lot more fun.

But there was another, greater consolation in the large collection of oblique and vertical photographs of Frobisher Bay, now sitting in the Hydrographic Office. The intricate shorelines of that multitude of bays and islands offered ideal material with which to experiment on rectification of obliques. The project which set out to chart a supposedly vital area had furnished instead a wealth of research material for photogrammetry. Perhaps in thus paving the way for better charting methods for the Pacific, the Crystal survey had made its chief contribution to the war effort.

All through the winter and spring of 1944 the quest for maximum service of photogrammetry to the forces in the Pacific led on through conference, persuasion, experiment, and planning. And then one day great news came that the Japanese had hastily abandoned a ship aground on the beach in Kwajalein atoll, and in her hold had been found a vast store of charts showing in minute detail nearly every Japanese-held island in all the Pacific. This was a windfall of major import to the Navy. One engineer in "Hydro" estimated that this priceless find would save countless lives of our military personnel. The great stream of aerial photographs pouring in from both Army and Navy photographic squadrons could now be used to check the Japanese charts and bring them up to date, and thus accurate charts could be furnished the fighting forces in a fraction of the time that would otherwise have been needed. This was indeed cause for great rejoicing, but as the extra tanks in the P-38 airplanes and the defeat of the U-boats had rendered the northern air route superfluous, so now this fortunate windfall made the utility of oblique photogrammetry look insignificant. The development that might have served our Navy much, two years earlier, was now too late to do more than help with the few areas of which our charts were still inadequate.

In September 1944, when our revised charts of Frobisher Bay had been printed and sent north to help the few cargo ships still serving that area, a hurry call came that offered a real challenge in another quarter. Near Norfolk, Virginia, was a small but important laboratory where a group of photogrammetrists were making rubber models of places soon to be attacked, to aid the amphibious forces to make the landing. Photographs were flown directly from the Pacific, analyzed, and the models were speedily made and flown back to the front with the utmost dispatch. Some of the photographs were high obliques, and the officers in charge of the work came to the Hydrographic Office for advice on how to use them. The chief photogrammetrist at "Hydro" disliked oblique photographs, and knowing that I liked them, referred these men to me. The result was that I was called to Norfolk one day and given some oblique pictures from which to locate two islands off the beach of a larger island south of the Philippines, where an important landing was soon to be made, and a 24-hour deadline for completing the task. Halfway through this exciting race against time, the great hurricane of 1944 blew down trees all around the laboratory and put out the lights, but by the dim daylight filtering through a grimy window I managed to place the islands, left my sketch to be incorporated in the model, and returned to Washington.

About the time this model was being made, an important change of plan occurred in the Pacific, and as I watched the dispatches for word of the expected landing, we were greeted by the heartening news of the successful occupation of Leyte several hundred miles nearer Tokyo. The model had become superfluous, like the northern air route and the photorectifier.

After this the tortuous trail that began at Labrador in 1931 led me to the Caribbean to use photogrammetry in an exhaustive study of smokescreens. This interesting research by a group of a dozen experts went on for six months and more, and an impressive contribution to the knowledge of this important adjunct of naval warfare was hurriedly flown to the Pacific in the closing days of July, 1945.

The phrase "too little and too late" was frequently applied to causes of military disaster. But it had a happier meaning when applied to things that became superfluous through successful achievement of the desired end by other means. It was my lot, I believe, to be involved in more of these fortunate futilities than any other man in the Navy. The crowning example was the relegation of the ponderous smoke-screen report to the files, when Japan surrendered before the knowledge could be applied. The consolation was easy this time.

Thus the trail led to many strange places and arenas where the talisman of oblique photogrammetry, revealed to me by Miller in 1931, might solve some urgent problem of war. Bright vistas opened of important mapping projects, quickly swallowed up in clouds of frustration, brewed by vapors of human frailty and by the shifting winds of fate.

At last, the war over, the trail led to the heart of the Crossroads project where from high in a Naval plane in July 1946 I looked down on Bikini Lagoon and watched while a mass little or no larger than a man's head released the energy that threw millions of tons of water high into the air, atomized the ship from which it was suspended, and sank a battleship before the great cloud of foam, spray, and fog had cleared away. But all that is another story.

BIBLIOGRAPHY

BIBLIOGRAPHY

[Anonymous], "Thunder over the North Atlantic," *Fortune* (November 1944), p. 153. (Believed to be the first published mention of the Northern Air Route.)

Bartlett, Robert A., "Servicing Arctic Airbases," illustrated by D. S. Turner and S. A. Wengerd, *The National Geographic Magazine* (May 1946).

Davidson, W., "Our Sub-Zero Heroes," *Collier's* (June 21, 1952), p. 28.

Dunbar, M. J., "The Ungava Bay Problem," *Arctic,* vol. V (March 1952).

Forbes, Alexander, "Surveying in Northern Labrador," *Geographical Review,* vol. XXII (1932), p. 30.

Forbes, Alexander, "A Flight to Cape Chidley," *Geographical Review,* vol. XVI (1936), p. 48.

Forbes, Alexander, *Northernmost Labrador Mapped from the Air* (American Geographical Society, 1938).

Forbes, Alexander, "Rivers of the South Shore of Lake Melville, Labrador," *Geographical Review,* vol. XXX (1940), p. 394.

Forbes, Alexander, "On the Northern Air Route," *Harvard Alumni Bulletin,* vol. LI (1949), p. 505.

Hall, Charles F., *Arctic Researches and Life among the Esquimaux* (Harper & Brothers, 1865).

Holdridge, Desmond D., "Down the Labrador," *Fore An' Aft* (January 1928).

Kane, Elisha K., *Arctic Explorations* (R. W. Bliss & Co., 1868).

La Farge, Oliver, *The Eagle in the Egg* (Houghton Mifflin Co., 1949).

Mirsky, Jeannette, *To the Arctic* (A. A. Knopf, 1948).

Morison, Samuel E., *History of United States Naval Operations in World War II,* vol. I (Little, Brown & Co., 1947), p. 62.

Pocock, Arthur, *Red Flannels and Green Ice* (Random House, 1949).

Streeter, Daniel W., *An Arctic Rodeo* (Garden City Publishing Co., 1929).

Wengerd, Sherman A., "Elevated Strandlines of Frobisher Bay, Baffin Island, Canadian Arctic," *Geographical Review,* vol. XLI (1951), p. 622.